KU-686-922

PC Audio Editing

Broadcast, desktop and CD audio production

Second Edition

Roger Derry M.I.B.S.

14784SX

CHESTER COLLEGE

ACC No. 01091864 | DEPT

CLASS No. 621·3893 DER

LIBRARY

WITHDRAWN

Focal Press

OXFORD AMSTERDAM BOSTON LONDON NEW YORK PARIS
SAN DIEGO SAN FRANCISCO SINGAPORE SYDNEY TOKYO

Focal Press
An imprint of Elsevier Science
Linacre House, Jordan Hill, Oxford OX2 8DP
200 Wheeler Road, Burlington MA 01803

First published 2000
Reprinted 2000
Second edition 2003

Copyright © 2003, Elsevier Science Ltd. All rights reserved

No part of this publication may be reproduced in any material form (including
photocopying or storing in any medium by electronic means and whether
or not transiently or incidentally to some other use of this publication) without
the written permission of the copyright holder except in accordance with the
provisions of the Copyright, Designs and Patents Act 1988 or under the terms of
a licence issued by the Copyright Licensing Agency Ltd, 90 Tottenham Court Road,
London, England W1T 4LP. Applications for the copyright holder's written
permission to reproduce any part of this publication should be addressed
to the publisher

British Library Cataloguing in Publication Data
A catalogue record for this book is available from the British Library

Library of Congress Cataloguing in Publication Data
A catalogue record for this book is available from the Library of Congress

ISBN 0 240 51697 4

For information on all Focal Press publications visit our website at:
www.focalpress.com

Composition by Genesis Typesetting, Rochester, Kent
Printed and bound in Great Britain

Contents

Contents

Foreword

Since the mid 1990s and into the new millennium, digital audio has evolved into a technology that actually comes close to its original vision of being a 'recording studio in a box'. That's to say, the digital audio workstation is (at its best) not only powerful, easy-to-use and cost-effective. . . but with the advent of faster processors, better programming and improved internal/external hardware devices, the DAW 'digital audio workstation' has finally become a master of system's integration. By this, I mean that new breeds of digital audio editors are not only capable of handling such tasks as:

- Recording, editing, ripping audio directly from CDs, importing and arranging loops
- Arranging, editing and overdubbing within a multitrack editing environment
- Processing and mixing audio in real-time and then mixing them down to any number of file formats

. . . but are also able to communicate with electronic instruments and processor via MIDI, synchronize a project to film or video (via an integrated video window) and even burn the finished music or audio project directly to CD . . . all without leaving the program! If this ain't integration heaven, I don't know what is!

One of the most recent advances for achieving software and hardware integration to hit the technological limelight is USB (Universal Serial Bus), which gives us affordable access to such devices as hardware controllers (devices that let us physically mix, control and vary edit/program parameters directly from a hardware control surface), audio interfaces (that gives us access to more and/or higher-quality audio inputs and even video capture devices (for importing video files)) . . . all without the need of a dedicated hardware card or external power source. In fact, hardware integration has actually come so far, that within a number of minutes, I can actually plug an external audio interface/controller and 2nd video LCD monitor into the laptop that I'm currently writing on . . . and begin the music-making process in a sophisticated production environment that still staggers my mind.

Within this book, Roger Derry introduces us to the wonderful world of audio recording, editing, processing, mixing and systems integration for the PC . . . This is done with the help of one of my favorite of all editing programs – Cool Edit Pro 2.

If you're like me, you'll really enjoy the straightforward simplicity, versatility, power and cost-effectiveness of this editor, which takes advantage of recent advances in audio production for the

PC. Finally . . . Let's get to the really important part of learning how to use the PC for audio editing . . . Spend time practicing on it! Play with it! Learn from it! Grow with it! and most of all . . . Have Fun with it!

David Miles Huber
www.modrec.com
www.51bpm.com

Preface

For years PCs struggled hard to cope with the prodigious demands of handling audio, let alone doing it well, or at any speed. Nowadays 'entry-level' PCs can provide facilities that, in previous decades, professional audio technicians would have killed for. Indeed, there is no doubt that a contemporary 6-year-old could happily edit audio on a PC. But, as always with most skills, the trick is NOT doing it like a 6-year-old child.

This book is aimed at people who wish to make audio productions for issue as recordings, or for broadcast, using a Windows PC with material acquired using portable equipment, as well as in studios. There are already plenty of books on music technology; this one is aimed at people who want to use speech as well – the making of cassette sales promotions, radio programmes or even *Son et Lumière*! The book does *not* expect you to be a technician, although if you are there will still be plenty to interest you, as the book includes production insights to help you use your PC for producing audio productions in the real world.

Even those who are experienced in editing and mixing with quarter-inch tape can find the change to editing audio visually on a PC daunting. However, while this does bring a change of skills, most of the new skills are more easily learned.

Visual editing is done statically rather than having to use the arcane dynamic skills of 'scrub' editing required by quarter-inch tape. Hearing an edit while scrubbing the tape back and forth was a skill that took weeks or even months to perfect. Some were doomed never to be able to hear clearly. With the PC, making mistakes is far less of a problem; correcting errors is rapid, and safe, unlike unpicking bits of tape spliced with sticky tape.

The author's early career included being a specialist tape editor. He is now a convert to the use of a PC for editing. His career has included all the jobs that go to make a radio programme, including production, reporting, management, studio and location technical operations (and fetching tea!). He can therefore offer advice and experience over the whole range of audio production.

This book is *not* a manual for a particular PC editor, although it is firmly based on the widely used *Cool Edit Pro*. This second edition uses version 2, which has been developed since the first edition was published. *Cool Edit 2*, a cheaper, reduced facility program, has also been produced since then. In its basic form it is a wave editor with reduced facilities. Extra modules can be bought to bring it up to a program that can do much of what is possible with the *Pro* version. There are, at the time of writing, some cosmetic differences between the two programs, but the operating techniques are very similar. This book has been written using a late 'beta' version of *Cool Edit Pro* version 2, and there may be minor changes in the published version as well as in any later revisions.

The book attempts to show the basic principles of the new technology. While each of the many audio editors that exist have their operational differences, basic principles remain constant. Many aspects of *Cool Edit Pro* are described in detail. Yet, audio is audio. Digits are digits. Other programs may implement any task differently, better or worse, but the principles behind those operations remain the same.

Since the first edition PCs have become even faster, and with this speed new facilities have become feasible. The most notable is the ability to use 'real-time' effects such as equalization and reverberation. *Cool Edit Pro* makes full use of this with almost all of its Effects transforms available in real time as well as for changing audio files permanently. It also features a useful 'half-way house' where real-time effects can be 'locked'. When this is done a preprocessed version of the track is created, thus reducing the amount of processor power required to play a multitrack recording.

1

Visual editing

Anyone who is used to editing audio using quarter-inch tape may approach editing on a PC with trepidation. PCs allow you to edit visually rather than using your ears to 'scrub' the audio back and forth to hear the edit point. Some Digital Audio Workstations provide a simulation of scrub editing, either using special hardware in the form of a search wheel or providing an option to use the computer's mouse. This option is provided by some PC and Mac software such as *Pro Tools*. These scrub options are often unsatisfactory because of tiny but noticeable processing delays between the mouse movements and hearing the result. As a professional editor I used to find that the delay caused by having loudspeakers more than 2 metres away was sufficient to cause problems, and that corresponds to only to 6/1000 of a second!

However, editing on a PC is a different medium, and techniques change. After 25 years of quarter-inch tape editing, I found that I took to editing visually like the proverbial duck to water. I did, however, have to overcome an emotional resentment at what seemed like a 'de-skilling' of the task.

My analogy is with word processing. In the days of manual typewriters and using carbon paper to make copies, there was a high premium on accuracy. It was also important to be able to type each key with an even pressure, so that the document had a professional look. This was a skilled thing to do, and was not learnt in a day. When word processors were introduced, the rules changed. It was soon discovered that it was more efficient to type as fast as possible and to clean up typos afterwards. Printouts take no notice of the key pressures that the typist has used. This means that even a person with no keyboard skills can – given a great deal of time – produce a document with a professional appearance.

So it is with audio editing; with plenty of experience and good training it is possible to find the precise edit point very quickly, mark, cut and splice, to get a good edit every time. It is a skill, and has to be learned.

The PC offers visual tools to aid your editing, and also has the potential to be much more accurate. With reel-to-reel quarter-inch analogue tape at 15 ips, using 60° cuts, the effective best accuracy is 1/120th of a second. Ninety-degree cuts can improve this, but, in practice, not by much. Most edits are made using much lower accuracy.

In contrast, the PC can offer 'sample-rate' accuracy. This means that you can edit down to the resolution of a single number (sample) in the digital data. Using the standard professional sampling rates of 32 kHz, 44.1 kHz or 48 kHz, this gives a possible accuracy of 1/32 000th, 1/44 100th or 1/48 000th of a second! As has already been observed, most edits don't need this accuracy, so it is as well that you are not forced to work to this resolution! Most editors will provide a crossfade across the edit. This corresponds to the crossfade provided by slanting the cut on quarter-inch tape.

When viewing the sound file as a whole, the PC offers you a static graphical display of your audio with level represented by the width of a line. For stereo there are two variable width lines side by side, representing the left and right channels. You can choose at what resolution you want to look at the audio. The width of the screen can encompass several hours, or one-thousandth of a second.

Figure 1.1 illustrates a 4-second chunk of mono audio. The text of what is being said has been added to the illustration. PC audio editors are not (yet) clever enough to transcribe speech from your recording!

Figure 1.1 A 4-second chunk of mono audio, with text added

At this sort of resolution, you can easily see the rises and falls in level corresponding to individual syllables. Yet we can zoom in to even more detail. The usual way to do this is to select the area by dragging the mouse – just as you would to select text in a word processor. You then zoom to the selection. In this example using the word 'and' (Figure 1.2) is enough to show the individual vibrations (Figure 1.3). The 'D' sound at the end is now extremely easy to see.

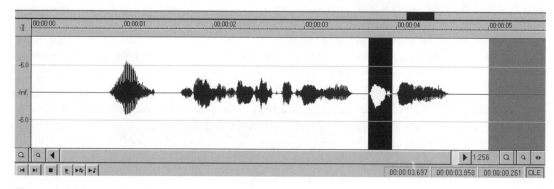

Figure 1.2 The word 'and' selected

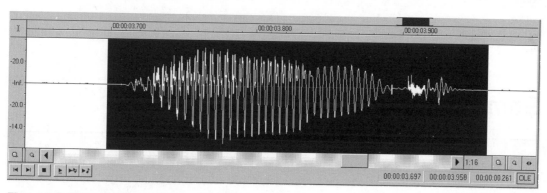

Figure 1.3 Zoomed in on the word 'and'

Cutting a section of audio is as easy as selecting it with a mouse and pressing the delete key, just as in word processing. Also, as in word processing, you can cut or copy audio to a clipboard and then paste it elsewhere.

There is usually an option for a 'mix' paste, where the copied audio is mixed on top of existing audio rather than being inserted. With several attempts and much use of the UNDO function this can be a way of adding a simple music bed to speech, but it lacks the sophisticated control available from using a non-linear editing mode to give you multitrack operation.

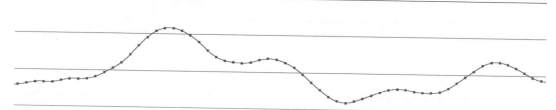

Figure 1.4 Audio zoomed in to showing samples as dots; the line joining them is created by the computer

Most audio editors allow you to zoom to the sample level (Figure 1.4), but this is rarely of much value except where you are manually taking out a single click from an LP transfer or some glitch picked up on the way. In general, noise reduction and de-clicking is done more quickly by software, although this will let through some clicks that have to be dealt with individually. Too high a setting on de-click software will 'overcook' the audio, introducing distortion where it starts attacking the sharp peaks of the audio.

2

Some technical bits

There are some technical terms that are much bandied about in audio. While it is possible to survive without a knowledge of them, they are a great help in making the most of the medium. However, you may wish to skip this chapter and read it later.

2.1 Loudness, decibels and frequencies

How good is the human ear?

Sound is the result of pressure – of compression/decompression waves travelling through the air. These pressure waves are caused by something vibrating. This may be something obvious, like the skin of a drum or the string and sounding board of a violin. However, wind instruments also vibrate; vibration may be caused by blowing through a reed or across a hole, and the turbulence causes the column of air within the pipe of the instrument to vibrate.

The two major properties that describe a sound are its frequency and its loudness.

Frequency

Frequency is a count of how many times per second the air pressure of the sound wave cycles from high pressure, through low pressure and back to high pressure again (Figure 2.1). This used to be known as the number of 'cycles per second', but has now been given a metric unit name; 1 hertz (abbreviation Hz) is one cycle per second. The hertz is named after Heinrich Hertz, who did fundamental research into wave theory in the nineteenth century.

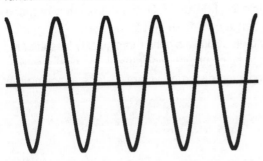

Figure 2.1 Five cycles of a pure audio sine wave

The lowest frequency the ear can handle is about 20 Hz. These low frequencies are more felt than heard. Some church organs have a 16-Hz stop, which is added to other notes to give them depth. Low frequencies are the hardest to reproduce, and in practice most loudspeakers have a tough time reproducing much below 80 Hz.

At the other extreme, the ear can handle frequencies of up to 20 000 Hz, usually written as 20 kHz (kilohertz). As we get older our high frequency limit reduces, and this can happen very rapidly if the ear is constantly exposed to high sound levels.

The standard specification for high-fidelity audio equipment is that it should handle frequencies between 20 Hz and 20 kHz equally well. Stereo FM broadcasting is restricted to 15 kHz, as is the NICAM system used for television stereo in the UK. However, digital radio is not, and so broadcasters have begun to increase the frequency range required when programmes are submitted.

The frequencies produced by musical instruments occupy the lower range of these frequencies. The standard tuning frequency 'middle A' is 440 Hz, the 'A' one octave below that is 220 Hz, and an octave above is 880 Hz. In other words, a difference of an octave is achieved by doubling or halving the frequency.

Instruments also produce 'harmonics'. These are frequencies that are multiples of the original 'fundamental' note. It is these frequencies, along with transients (how the note starts and finishes), that give an instrument its characteristic sound. This is often referred to as the timbre (pronounced 'tam-ber').

Loudness

Our ears can handle a very wide range of levels. The power ratio between the quietest sound that we can just detect – in a quiet, sound-insulated room – and the loudest sound that causes us pain is:

$$1 : 1\,000\,000\,000\,000$$

'1' followed by twelve noughts is one million million! To be able to handle such large numbers a logarithmic system is used. The unit, called a bel, can be thought of as a measure of the number of noughts after the '1'. In other words, the ratio shown above could also be described as 12 bels. Similarly, a ratio of 1 : 1000 is 3 bels, and a ratio of 1 : 1 – no change – is 0 bels. Decreases in level are described as negative, so a ratio of 1000 : 1 – a reduction in power of 1/1000th – is minus 3 bels.

For most purposes, the bel is too large a unit to be convenient. Instead, the unit used every day is one-tenth of a bel. The metric system term for one-tenth is 'deci', so the unit is called the 'decibel'.

Table 2.1 Everyday sound levels

0 dB	Threshold of hearing: sound-insulated room
10 dB	Very faint – a still night in the country
30 dB	Faint – public library, whisper, rustle of paper
50 dB	Moderate – quiet office, average house
70 dB	Loud – noisy office, transistor radio at full volume
90 dB	Very loud – busy street
110 dB	Deafening – pneumatic drill, thunder, gunfire
120 dB	Threshold of pain

The abbreviation is 'dB' – little 'd' for 'deci' and big 'B' for 'bel', as it is based on a person's name; in this case Alexander Graham Bell, the inventor of the telephone and founder of Bell Telephones, who devised the unit for measuring telephone signals. Conveniently, a change of level of 1 dB is about the smallest change that the average person can hear (Table 2.1); also

- 3 dB represents a doubling of power; 6 dB represents a doubling of voltage
- 10 dB increases sounds to twice as loud; –10 dB decreases sounds to be half as loud.

(Power equals Voltage × Current; double the voltage and you also double the current.)

2.2 Hearing safety

One of the hazards of audio editing on a PC is that it is often done on headphones in a room containing other people. Research has shown that people, on average, listen to headphones at 6 dB louder than they would to loudspeakers. So already they are pumping four times more power into their ears. When you are editing, there are going to be occasions when you turn up the volume to hear quiet passages and then forget to restore it when going on to a loud section. The resulting level into your ears is going to be way above that which is safe.

Some broadcasting organizations insist that their staff use headphones with built-in limiters to prevent hearing damage. By UK law, a sound level of 85 dBA is defined as the first action level (the 'A' indicates a common way of measuring sound-in-air level, as opposed to electrical audio decibels). You should not be exposed to sound at or above this level for more than 8 hours a day. If you are, as well as taking other measures, the employer must offer you hearing protection.

The second action level is 90 dBA, and at this noise level or higher ear protection must be worn, and the employer must ensure that adequate training is provided and that measures are taken to reduce noise levels as far as is reasonably practicable.

The irony here is that the headphones could be acting as hearing protectors for sound from outside, but themselves be generating audio levels above health and safety limits.

If you ever experience 'ringing in the ears' or are temporarily deafened by a loud noise, then you have permanently damaged your hearing. This damage will usually be *very* slight each time, but accumulates over months and years; the louder the sound, the more damage is done.

Slowly entering a world of silence may seem not so terrible, but if your job involves audio it will mean losing that job. Deafness cuts you off from people, and is often mistaken for stupidity. Worse, hearing damage does not necessarily create a silent world for the victim. Suicides have been caused by the other result of hearing damage, which is described by the medical profession as the gentle-sounding word 'tinnitus'. This conceals the horror of living with loud, throbbing sounds created within your ear. They can seem so loud that they make sleep difficult. Some people end up in a no-win situation, where to sleep they have to listen to music on headphones at high level to drown the tinnitus. Of course, this in turn causes more hearing damage.

If you are reading this book then you value your ears – so please take care of them!

2.3 Analogue and digital audio

Analogue audio signals consist of the variation of sound pressure level with time being mimicked by the analogous change in strength of an electrical voltage, a magnetic field, the deviation of a groove, etc.

The principle of digital audio is very simple, and that is to represent the sound pressure level variation by a stream of numbers. These numbers are represented by pulses. The major advantage of using pulses is that the system merely has to distinguish between pulse and no-pulse states. Any noise will be ignored provided it is not sufficient to prevent that distinction (Figure 2.2).

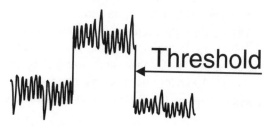

Figure 2.2 Distinction between pulse and no-pulse states

The actual information is usually sent by using the length of the pulse; it is either short or long. This means that the signal has plenty of leeway in both the amplitude of the signal (how tall it is) and the length of the pulse. Well-designed digital signals are very robust, and can traverse quite hostile environments without degradation. However, this can be a disadvantage for, say, a broadcaster with a 'live' circuit, as there can be little or no warning of a deteriorating signal. Typically the quality remains audibly fine until there are a couple of splats, or mutes, then silence. Analogue has the advantage that you can hear a problem developing and make arrangements for a replacement before it becomes unusable.

Wow and flutter are eliminated from digital recording systems, along with analogue artefacts such as frequency response and level changes. If the audio is copied as digital data, then it is a simple matter of copying numbers and the recording may be 'cloned' many times without any degradation. (This applies to pure digital encoding; however, many modern systems, such as Minidisc and Digital Radio and Television, use a 'lossy' form of encoding. This throws away data in a way that the ear will not usually notice. However, multi-generation copies can deteriorate to unusability within six generations, especially if different forms of lossy compression are encountered. Even audio CD and DAT are lossy to some extent, as they tolerate and conceal digital errors.)

What is digital audio?

Pulse and digital systems are well established, and can be considered to have started in Victorian times.

Perhaps the best known pulse system is the Morse code. Like much digital audio, this code uses short and long pulses. In Morse, these are combined with short, medium and long spaces between pulses to convey the information. The original intention was to use mechanical devices to decode the signal, but in practice it was found that human operators could decode by ear faster.

Modern digital systems run far too fast for human decoding, and adopt simple techniques that can be dealt with by microprocessors with rather less intelligence than a telegraph operator. Most systems use two states that can be thought of as on or off, short or long, dot or dash (some digital circuits and broadcast systems can use three or even four states, but these systems are beyond the scope of this book).

The numbers that convey the instantaneous value of a digital audio signal are conveyed by groups of pulses (bits) formed into a digital 'word'. The number of these pulses in the word sets the number of discrete levels that can be coded.

The word length becomes a measure of the resolution of the system and, with digital audio, the fidelity of the reproduction. The compact disc uses 16-bit words giving 65 536 states. NICAM stereo, as used by UK television, uses just 10 bits, but technical trickery gives a performance similar to 14-bit. Audio files used on the Internet are often only 8-bit in resolution, while systems used for telephone answering etc. may only be 4-bit systems or less (Table 2.2).

Table 2.2 Table showing how many discrete levels can be handled by different digital audio resolutions

1-bit = 2	7-bit = 128	13-bit = 9192	19-bit = 524 287
2-bit = 4	8-bit = 256	14-bit = 16 384	20-bit = 1 048 575
3-bit = 8	9-bit = 512	15-bit = 32 768	21-bit = 2 097 151
4-bit = 16	10-bit = 1024	16-bit = 65 536	22-bit = 4 194 303
5-bit = 32	11-bit = 2048	17-bit = 131 071	23-bit = 8 388 607
6 bit = 64	12-bit = 4096	18-bit = 262 143	24-bit = 16 777 215

The larger the number, the more space is taken up on a computer hard disk or the longer a file takes to copy from disk to disk or through a modem. Common resolutions are: 8-bit, Internet; 10-bit, NICAM; 16-bit, compact disc; 18–24-bit, enhanced audio systems for production. Often, 32-bit is used internally for processing to maintain the original bit-resolution when production is finished.

Sampling rate

A major design parameter of a digital system is how often the analogue quantity needs to be measured to give accurate results. The changing quantity is sampled and measured at a defined rate.

If the sampling rate is too slow, then significant events may be missed. Audio should be sampled at a rate that is at least twice the highest frequency that needs to be produced. This is so that, at the very least, one number can describe the positive transition and the other the negative transition of a single cycle of audio. This is called the Nyquist limit after Harry Nyquist of Bell Telephone Laboratories, who developed the theory.

For practical purposes, a 10 per cent margin should be allowed. This means that the sampling rate figure should be 2.2 times the highest frequency, and as 20 kHz is regarded as the highest frequency that most people can hear, this led to the CD being given a sampling rate of 44 100 samples a second (44.1 kHz). The odd 100 samples per second is a technical kludge. Originally digital signals could only be recorded on videotape machines, and 44.1 kHz will fit into either American or European television formats.

The sampling process must be protected from out-of-range frequencies. These 'beat' with the sampling frequency and produce spurious frequencies that not only represent distortion but also, because of their non-musical relationship to the intended signal, represent a particularly nasty form of distortion. These extra frequencies are called alias frequencies, and the filters called anti-aliasing filters. It is the design of these filters that can make the greatest difference between the perceived quality of analogue to digital converters. You will sometimes see references to 'over-sampling'; this technique emulates a faster sampling rate (4×, 8×, etc.), and simplifies the design of the filters.

Errors

A practical digital audio system has to cope with the introduction of errors, owing to noise and mechanical imperfections, of recording and transmission. These can cause distortion, clicks, bangs and dropouts when the wrong number is received.

Digital systems incorporate extra 'redundant' bits. This redundancy is used to provide extra information to allow the system to detect, conceal or even correct the errors. Decoding software is able to apply arithmetic to the data, in real time, so it is possible to use coding systems that can actually detect errors.

However, audio is not like accountancy, and the occasional error can be accepted, provided it is in a well-designed system where it will not be audible. This allows a simpler system (using less redundant bits) to be used to increase the capacity, and hence the recording length, of the recording medium. This is why a CDR burnt as a computer CD-ROM storing wave file data has a lower capacity than the same CDR using the same files as CD-audio. CD-ROMs have to use a more robust error correction system, as NO errors can be allowed. As a result, a standard CDR can record 720 Mbytes of CD audio (74 minutes) but only 650 Mbytes of computer data.

Having detected an error, a CD player may be able to:

1 Correct the error (using the extra 'redundant' information in the signal)
2 Conceal the error, which is usually done either by sending the last correctly received sample (replacement) or by interpolation, where an intermediate value is calculated
3 Mute the error – a mute is usually preferable to a click.

While the second and third options are good enough for the end product used by the consumer, you need to avoid the build-up of errors during the production of the recording. Copies made on hard disk, internally within the computer, will be error-free. Similarly, copies made to CD-ROM or to removable hard disk cartridge will also have no errors (unless the disk fails altogether because of damage).

Multi-generation copies made through the analogue sockets of your sound card will lose quality. Copies made digitally to DAT will be better, but still accumulate errors. However, a computer backup as data to DAT (4 mm) should be error-free (as should any other form of computer backup medium. These have to be good enough for accountants, and therefore error-free).

How can a computer correct errors?

It can be quite puzzling that computers can get things wrong but then correct them. How is this possible? The first thing to realize is that each datum bit can only be '0' or '1'. Therefore, if you can identify that a particular bit is wrong, you know the correct value – if '0' is wrong then '1' is right, and if '1' is wrong then '0' is right.

The whole subject of error correction involves deep mathematics, but it is possible to give an insight into the fundamentals of how it works. The major weapon is a concept called parity. The basic idea is very simple but, suitably used, can become very powerful. At its simplest this consists of adding an extra bit to each data word, and this bit signals whether the number of '1's in the binary data is odd or even. Both odd and even parity conventions are used. With an even parity convention the parity bit is set so that the number of '1's is an even number (zero is an even number), and with odd parity the extra bit is set to make the number of '1's always odd.

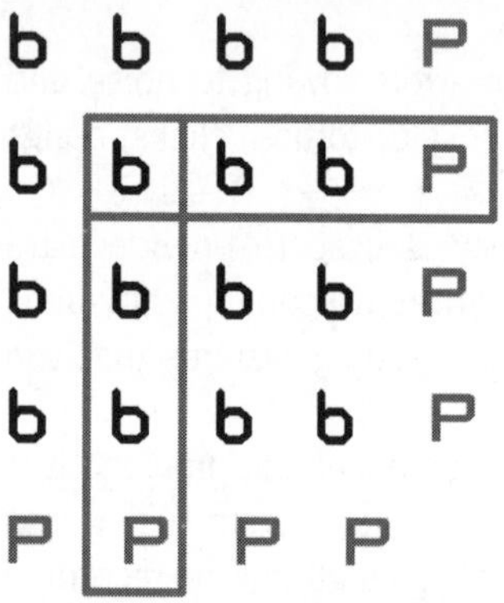

Received data are checked during decoding. If the signal is encoded with odd parity and arrives at the decoder with even parity, then it is assumed that the signal has been corrupted. A single parity bit can only detect an odd number of errors.

Quite complicated parity schemes can be arranged to allow identification of which bit is in error, and for correction to be applied automatically. Remember that the parity bit itself can be affected by noise.

Figure 2.3 illustrates a method of parity checking a 16-bit word by using eight extra parity bits. The data bits are shown as 'b', the parity bits are shown as 'P'.

Figure 2.3 A method of parity checking a 16-bit word by using 8 extra parity bits

The parity is assessed both 'vertically' and 'horizontally'. The data are sent in the normal way, with the data and parity bits intermingled. This is called a Hamming code. If the bit in the second column, row two were in error, its 2 associated parity bits would indicate this. As there are only two states, if the bit is shown to be in error then reversing its state must correct the error.

Dither

The granular nature of digital audio can become very obvious on low-level sounds such as the die-away of reverberation or piano notes. This is because there are very few numbers available to describe the sound, and so the steps between levels are relatively larger as a proportion of the signal.

This granularity can be removed by adding random noise, similar to hiss, to the signal. The level of the noise is set to correspond to the 'bottom bit' of the digital word. The frequency distribution of the hiss is often tailored to optimize the result, giving noise levels much lower than would be otherwise expected. This is called 'noise-shaping'. PC audio editors often have an option to turn this off, but don't do this unless you know what you are doing. Dither has the almost magical ability to enable a digital signal to carry sounds that are quieter than the equivalent of just 1 bit.

2.4 Time code

All modern audio systems have a time code option. With digital systems, it is effectively built in. At its simplest level, it is easy to understand; it stores time in hours, minutes and seconds. As is so often the case, there are several standards.

The most common audio time display that people meet is on the compact disc; this gives minutes and seconds. For professional players, this can be resolved down to fractions of second by counting the data blocks. These are conventionally called frames, and there are 75 every second. This is potentially confusing, as CDs were originally mastered from three-quarter-inch U-Matic videotapes where the data were configured to look like an American television picture running at 30 video frames per second (fps).

In 1967, the Society of Motion Picture and Television Engineers (SMPTE) created a standard defining the nature of the recorded signal and the format of the data recorded. This was for use with videotape editing. Data are separated into 80-bit blocks, each corresponding to a single video frame. The way that the data are recorded (Biphase modulation) allow them also to be read from analogue machines when the machines are spooling at medium speed, with the tape against the head, in either direction. With digital systems, the *recording* method is different but the *code* produced stays at the original standards.

There are three common video frame standards; 25 fps (European TV), 30 fps (American) and, for technical reasons, a more complicated format known as 30 fps drop frame, which corresponds to an average to 29.97 fps.

By convention, on analogue machines the highest numbered track is used; track 4 on a 4-track; track 16 on a 16-track, etc. It is a nasty screeching noise best kept as far away from other audio as possible.

Time code can also be sent to a sequencer (via a converter) as MIDI data, allowing the sequencer to track the audio tape. The simple relationship between bars, tempo and SMPTE time as shown by sequencers like Cubase is only valid for 120 beats per minute 4/4 time. MIDI time code generators need to be programmed with the music tempo and time signature used by the sequencer, so they can operate (in a gearbox fashion) so that the sequencer runs at the proper tempo.

People editing audio for compact disc will often prefer to set the time display to 75 fps to match the CD data. There is a small technical advantage to ensuring that an audio file intended for CD ends exactly at the frame boundary, as there are occasions when not doing so will cause a click.

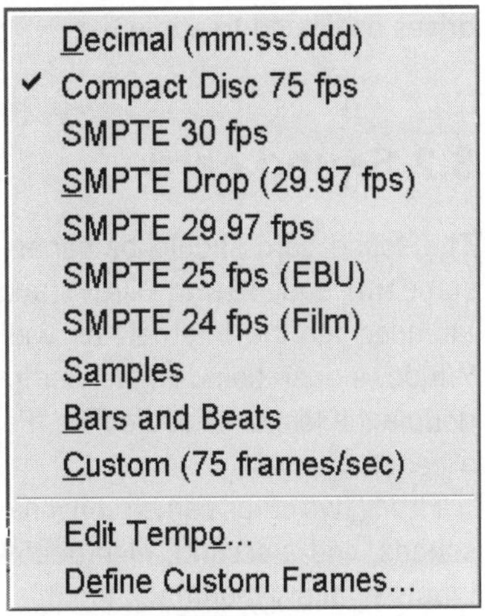

| Decimal (mm:ss.ddd) |
| ✓ Compact Disc 75 fps |
| SMPTE 30 fps |
| SMPTE Drop (29.97 fps) |
| SMPTE 29.97 fps |
| SMPTE 25 fps (EBU) |
| SMPTE 24 fps (Film) |
| Samples |
| Bars and Beats |
| Custom (75 frames/sec) |
| Edit Tempo... |
| Define Custom Frames... |

Figure 2.4 Right clicking on *Cool Edit*'s time display provides a wide range of options for the frame display

Right-clicking on *Cool Edit*'s time display provides a pop-up with a wide range of options for the frame display (Figure 2.4), including being able to match to the bars and beats of a MIDI track.

See Appendix 1 for more information on time code.

3

Hardware and software requirements

3.1 PC

Audio editing on a PC became a practical proposition for professional use once the equivalent of (or better than) a Pentium II running at more than 200 MHz was reached. These days this is regarded as being very slow, but even so will give more than adequate speed of processing and is able to perform multitrack mixing.

Consider getting a 'full tower' case as this will conveniently sit on the floor, releasing space on your desk. Floor-standing towers can also be strapped down more easily if theft is likely to be an issue. Tower cases have plenty of room for adding extra drives, and these can mount up very quickly. A reasonably full quota might be a CD-ROM drive, a CD-RW drive, a DVD drive, a removable hard disk system, two IDE drives to separate programs and system from data, and perhaps two SCSI drives dedicated to audio use.

3.2 Sound card

The sound card should be capable of at least 16-bit audio at sampling rates of 44.1 kHz (CD standard) and 48 kHz (DAT standard). Lower rates will be useful if you need to audition files intended for the Internet, as well as for reproducing noises made by other programs and by Windows operations. Higher sampling rates (88.2 kHz and 96 kHz have become fashionable, and double the length of audio files. People still argue a great deal as to whether this actually provides a better sound.

Fitting two cards can be convenient. A basic games type card will handle the Windows and games sounds, and a second, high quality one can be used for your audio editing – maybe with several inputs to allow multitrack input.

If you have access to high quality analogue sources such as 15/30 ips Dolby SR recordings or high quality microphones in quiet studios, then a higher bit rate is extremely desirable, even if your finished recording is going to be 16-bit. This is because headroom will have to be allowed at the time of recording to prevent digital overload. Quite likely a few levels will have been increased during post-production, and this means that you can end up delivering the equivalent of a 12-bit recording! (See section 5.4, Analogue.)

Some sound cards have microphone inputs, but these are usually of relatively poor quality. They are best faded out on the Window mixer to prevent them adding noise. The useful inputs are the 'line' level ones, which will match the line outputs of audio gear. Multitrack cards often have the option of handling professional levels, which are 14 dB higher than domestic levels. They may also be able to handle 'balanced' feeds as well as the domestic-style unbalanced type (see page 40).

As well as conventional analogue inputs and outputs, it can be useful if your sound card will handle digital inputs and outputs so your original digital recordings can be 'cloned' on to your computer and your final mix returned without quality loss.

In practice, many of the domestic machines – often used by many professionals – have only optical digital outputs rather than the electrical outputs required by many sound cards. However, you can buy converter boxes that will do the job very adequately but at the expense of yet another box cluttering up your desk.

Optical connectors use modulated red light rather than electricity to transfer the data. This light is visible to the naked eye when a connector is carrying an output. Be aware that there two physically different optical standards; one is called TOSLINK, and the other is the same shape and size as minijack audio connectors. The socket is usually dual function and can be used for electrical analogue audio or for optical digital (see page 40).

Your card should be switchable between the domestic and professional data formats. While they are nominally compatible, there can be circumstances where a domestic format input can interpret a professional style output as being copy protected.

Sound cards are traditionally slotted into the back of the computer, after removing the cover. This is probably as good a place as any; the card is kept safe and is not likely to be dropped. However, various boxes that plug into the Universal Serial Bus or Firewire input of the computer can be used instead, and these have the advantage that they can be plugged and unplugged without switching off the computer (see page 41). While some versions of Windows 95 claim to be able to use USB, you really need Windows 98 or later for success.

Most editing is done away from the recording site and the audio is brought to the computer on removable media. However, you may need one or more microphones. For simple one-mic use for voice-overs or story readings, there are plenty of devices that provide a clean amplifier for a microphone – including phantom power if required. They often include equalization and compression as well. Some sound cards with breakout boxes have high quality mic inputs – usually about four will be supplied with a software mixer, provided if you don't want to use *Cool Edit* directly. Once you want more than four microphones you are really talking about a studio set up, and a proper mixer will not only be more convenient but will also provide useful things like talkback and auxiliary feed to the artist's headphones.

3.3 Loudspeakers/headphones

Ordinary PC 'games' loudspeakers are not adequate to assess sound quality. You should pay the extra for music quality active speakers ('active' means that the speakers have their amplifiers built in for driving directly from the sound card output). If you have the space, this can be a proper hi-fi system with the computer's sound card feeding a line level 'aux' input.

If you are editing audio at work, then much of the time you may be operating in a shared office using headphones. These should be of the best quality with a decent bass response. Try always to check mixes on loudspeakers if at all possible, as headphones give a very different impression to loudspeakers. As a generalization, a mix that sounds good on loudspeakers will sound good on headphones. The reverse is not always true. It is all too easy to have too little separation between a voice and music or effects behind it.

Headphones are much more critical of poor editing than loudspeakers. The standard advice to broadcasters used to be not to worry if an edit was audible on headphones but not on loudspeakers. However, with the popularity of Walkman-style cassette and Minidisc players with radios, it is no longer possible to assume that all your audience is listening on loudspeakers.

Sometimes the audience is known to be listening on headphones – as in walk-round cassette guides to exhibitions, for example. Here, headphones will give you accurate results. Use your best headphones for production, but do check what the mix sounds like on the (probably cheaper) headphones used at the show.

3.4 Hard disks

Ideally you should have a second separate hard drive for your audio data – or even better, for the fastest operation, yet another separate disk for handling the audio editor's temporary files. This reduces head clacking where the hard disk head is flipping between the source file and the destination file during copying. This is eliminated if the source file is on a different drive from the destination file.

Traditionally the advice was that these drives should be of audiovisual (AV) quality, especially if you intended to 'burn' compact discs. Early, non-AV quality disks recalibrated themselves at inconvenient times, to compensate for temperature changes, interrupting a continuous flow of audio (or video). This is only likely to be a problem if you are intending to use very old equipment; modern drives should have no difficulty in delivering the data when required, but as always you should confirm with your supplier that the drives supplied are suitable for the task.

Modern IDE/EIDE drives are entirely adequate for audio purposes, although many people still think that the SCSI system is well worth the extra cost for AV work. SCSI also has the advantage of being able to handle up 15 devices (which can include scanners as well as hard drives) rather than the four IDE drives that most computers support.

If you are likely to be working on several projects at once, then a removable hard disk cartridge system will be useful. Cartridges of 1 Gbyte or greater can hold enough audio to make a CD – including spare material and auxiliary files. These are more expensive than the commonly available 100- and 250-Mbyte cartridges; however, the lower capacity media are able to take 10 or 25 minutes of CD quality stereo audio, which is entirely adequate for short items.

While less convenient, another alternative is to use standard hard drives mounted in removable caddies. This way, tens of gigabytes of data can be carried from computer to computer, provided that the same caddy system is used in each computer. The cheapest of these require you to turn off the computer to change caddies, but systems are available that allow 'hot' swapping and will update the Windows drive list on-the-fly. The caddies are more expensive than they ought to be, as most people need more drive carriers than slots to put them in, but it is difficult to buy the two halves of the system separately.

3.5 Universal serial bus

The universal serial bus (USB) has become very common, and there is a great deal of equipment that can use it. From the user's point of view it has the major advantage that it is not necessary to open up the computer to connect devices. Even better, USBs can be connected and disconnected without switching off or rebooting the computer. The disadvantage can be that you end up with a clutter of devices on flying leads scattered around it.

As well as things like CD burners, sound cards, digital cameras, scanners and external drives, basic things such as a mouse, a keyboard and a printer can be added. Theoretically, over a hundred separate devices could be hung on to a single USB port, daisy-chained one after the other. In practice, rather less than this can be used. The amount of data that can be sent at any time is limited, and although this is still vastly more than the ordinary serial port it will limit data-hungry devices. A CD burner will probably be limited to ×8 speed, whereas SCSI and IDE connections can be triple that. USB scanners are slower than SCSI ones.

USB2 connections are faster, but are still limited. If you are going to use USB devices, then you may consider purchasing a USB hub so that devices can have their own individual sockets rather than having to be daisy-chained. This prevents having to disconnect other devices when removing one in the middle of the chain. In fact, most devices do not have the necessary socket to connect to the next device in the chain anyway. Hubs can come as separate boxes or, these days, many monitors have hubs built in. USB keyboards usually have chaining sockets to take a USB mouse plus, say, a Zip drive.

Many modern computer motherboards have USB ports built in, but for those that don't PCI cards containing two or more ports can be added to an existing computer.

3.6 Firewire

Firewire is an alternative to USB. It is much faster, but at the time of writing there are less products on the market for it. Again PCI cards can be bought to add ports to existing computers. There is no reason why both USB and Firewire cannot be used on the same computer.

3.7 Audio editors

The audio editor is the key to the whole operation. There are any number of different ones made by a host of manufacturers. These range from full-scale Digital Audio Workstations (DAWs), often using proprietary hardware, to very simple 'freebies' with the minimum of facilities.

Most of the illustrations in this book will be taken from a good example of presently available software. *Cool Edit Pro* is a combined linear/non-linear editor potentially able to handle 128 audio tracks, and a good all-round general workhorse. Its methods are easily transferred to other editors. It also has a sister program, *Cool Edit 2000*, which has a subset of the *Pro* version's facilities. These can be enhanced by add-on modules.

3.8 Linear editors

Editors fall into two methods of working: linear and non-linear. Linear editors can handle one stereo, or mono, audio track at a time. They operate like word processors; any editing is destructive in that a deleted section is actually deleted from the audio file (although, as with word processors, there is likely to be an 'undo' feature available). When saved, any cut material is lost unless a backup copy of the original has been kept.

Owing to the large amounts of data involved with audio, cut and paste are instantaneous only for the smallest of sections. Copying a 40-minute stereo file can take several minutes. While the process may be much quicker than with pre-digital technology, minutes spent staring at a progress bar crossing your screen are frustrating and unproductive (see section 7.2, Blue bar blues).

3.9 Non-linear editors

Non-linear editors do not change the audio files being edited in any way, but instead create 'Edit Decision Lists' (EDLs). They don't play an audio file linearly from beginning to end; instead the files are played out of sequence – non-linearly – with edits performed by skipping instantly to the next section. It is a sophisticated version of programming the playback of a CD, where you set the CD only to play certain tracks but the omitted tracks are still on the disk. EDLs allow the playback to skip from instant to instant within and between tracks.

The most obvious advantage is that material is never lost. However, this can be a disadvantage as well, because your hard disk can soon become cluttered with unwanted material. In practice, some culling with a linear editor is useful before using a non-linear one.

You can have any number of different EDLs for the same audio, and thus any number of different versions. You can decide to keep a version that you are reasonably happy with, but continue to see if you can further improve the item. This is ideal if the same item has to be repackaged for different programme outlets. A programme trail can be one edit decision list, and the broadcast programme another. Even in the domestic environment different versions can sometimes be useful, even if it is just a shorter version for your mother to listen to!

While this is possible with linear editors, it is at the expense of much duplication of audio data, which can soon fill a hard disk. Only the non-linear editor can give you the opportunity to 'unpick' edits for your new version.

Non-linear editors can also handle more than one audio file at a time, and play from different sections of a single audio file simultaneously. In this way crossfades and mixes can be performed without permanently changing the original audio material.

In general, non-linear editing operations can be faster as the computer is not having to move multimegabyte chunks of digital audio data around the hard disk. Instead, the changes in the EDL are measured in just tens of bytes.

3.10 Multitrack

You don't have to handle complex EDLs yourself, as the non-linear editor presents itself, on screen, as if it were a multitrack tape recorder. The lists are managed by the editing software, and not by you. However, it is far more flexible than a physical multitrack tape recorder.

The non-linear editor shares, with multitrack tape, the ability separately to record individual tracks in synchronization with existing ones. You can also 'drop-in' to replace a section. However, unlike hardware multitrack with its physical restraints, you can also 'slide' tracks back and forth with respect to each other. Track 'bouncing' (copying from one track to another) is virtually instantaneous. You can add any level changes and crossfades in a way that would need full-scale automation in a studio using physical multitrack. When you have finished, you can usually get the editor to save space by performing the virtual edits on the audio files so that they become physical edits.

Just as over the years automobiles have become more and more like each other, so have multitrack audio editors. As an example, Figure 3.1 provides screen shots (at 800×600) of the same four audio files loaded into *Cool Edit Pro* on a PC, and *Cubase* and *ProTools* on Macintosh machines. A major difference between *Cool Edit* and the others is that it uses a single window for most of its actions. *Cubase* and *ProTools* tend to have separate windows. Most notably, the transport control is on a floating window, unlike the fixed control on *Cool Edit*. As an operator, moving from one to another is rather like driving different makes of cars. You may occasionally wash your windows at someone rather than flashing your headlights, and you may need to read the manual to be able to tune the radio, but basic driving – the main function – is relatively easily achieved.

3.11 Audio processing

The standard tricks of an audio editor are cut and paste, as in a word processor, plus the ability to crossfade between two pieces of audio.

In addition, audio editors for professional and semi-professional use come with a barrage of special effects. Some you will have daily use for, whereas others may 'get you out of a jam' some time. Many editor programs have the facility for their user to buy 'plugins' that will add specialist (or enhanced quality) effects. Some of the most sophisticated of these can cost more than the editor they plug in to.

The most important processing options are:

- *Normalization*. This standardizes the sound level of each item, although not necessarily the loudness.
- *Reverberation*. Sometimes known as 'artificial echo', this will add a room (or hall) acoustic to your recording. This is most often used with music recording. With speech programmes you are less likely to need it, except to get you out of a jam by recreating an acoustic. There is also the annual ritual on radio of adding reverberation to witches' voices for Halloween.

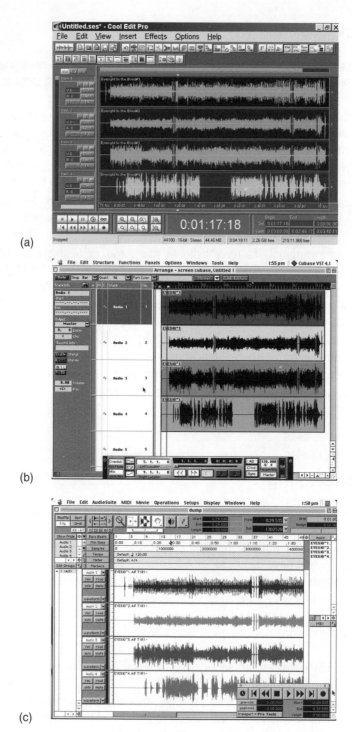

Figure 3.1 Screen shots of the same four audio files loaded into (a) *Cool Edit Pro*, (b) *Cubase*, and (c) *Pro Tools*

- *Compression*. This reduces the range of volume between the quietest and loudest sound. Used with care, it can enhance your recording. Used carelessly, it can make your item offensively difficult to hear. The overall effect is to make your material louder while remaining within the constraints of the maximum digital level available. It has the disadvantage of bringing up background noise and making voices breathy.
- *Noise reduction*. This should not be confused with the noise reduction used on analogue tape systems like dbx, Dolby A, B, C, S or SR. These are 'companding' systems, where programme material is compressed for recording but expanded back to its original dynamic range on playback. (Dolby A and Dolby SR are used in professional studios; Dolby B, C and S are domestic systems.) A digital system does not need such techniques. However, you may have a digital recording copied from a poor original, either one from the archive or one afflicted with technical problems at the time of the original recording. While intended to remove such recording problems as tape hiss, your editor's noise reduction system may well turn out to be useful in removing acoustic noise such as traffic rumble or air-conditioning whine. I have even managed to remove aircraft noises and telephones ringing.
- *Declicking*. Old 78 rpm recordings and vinyl LPs are plagued by clicks as well as noise, and will benefit from electronic declicking. However, don't be too enthusiastic about cleaning up old recordings, as they often *need* to sound old in the context of a programme item.
- *Filters*. These are useful for filtering out constant sounds as well as sound at the extreme ends of the frequency range. Cutting the bass will remove rumbles coming through the wall, or from underground trains. Mains hum can be a problem, and special 'notch' filters are available to remove the worst of this. Different settings are needed for American (60 Hz) and European (50 Hz) mains systems.
- *Equalization*. This is a sophisticated form of 'tone' control usually referred to as 'EQ'. It can be used to brighten up muffled recordings, or to reduce sibilance from some interviewees. Domestic tone controls normally only affect the top and bottom frequencies; equalizers can also do useful things to the middle frequencies.

3.12 Mastering

Mastering refers to the process of creating a recording that is usable away from the PC. It is possible to play (or broadcast) your mix from the computer directly to an audience. However, this does tie up your computer, with all its expensive resources. Therefore, you need to put your final mix onto a 'removable medium' in a conventional stereo form. This also has the not inconsiderable advantage that no one else can 'muck about with' your item.

Mastering can be as simple as playing the mix out of the computer's sound card onto a normal domestic medium like a compact cassette. This format has the major advantage of being playable almost universally, as the compact cassette is the most widely available sound recording medium yet devised.

Professionals may copy to DAT (preferably using a digital link) or to quarter-inch tape. Both these operations have to be done in 'real time' – 30 minute programmes take half an hour to dub.

In a broadcast environment, even now, dubbing to quarter-inch has the advantage that if an edit has been missed or a retake forgotten, then a quick, old-fashioned razor-blade cut can be made. This is quicker than correcting the edit on the computer and then having to start copying to DAT from the beginning.

3.13 CD recording software

The compact disc has become so widespread that it becomes a logical medium for distributing your item. Commercial CDs are pressed, not recorded individually, and for a long time this route was not practical.

The advent of the recordable CD (CDR) has therefore been a boon to all creators of audio material. However, it is still constrained by a number of factors:

1 Some early ordinary audio CD players may not recognize a CDR. This, in fact, is relatively unusual, and is unlikely to be a problem in a professional environment.
2 There are many different types of CD: CD-ROM, multi-session CD, etc. Domestic CD audio players (and many professional ones as well) only recognize one CD format – the original audio format. A consequence of this is that any compact disc that is to be used on a standard audio player needs to be recorded in a tightly defined way (as the process involves a laser making physical changes to the disc, it is usual to talk of 'burning' a CD).
3 There are also stand-alone CD recorders that take an audio or digital audio signal and record discs in real time. Many of these will only use so-called 'copyright paid' consumer blanks, which cost up to six or eight times as much as computer data blanks.
4 You can also buy 'rewritable' CDs that can be erased when the data they contain is finished with. They can be recorded as standard audio CDs, but while most computer CD-ROM drives and DVD players can play them, very many audio CD players cannot.
5 Most CD writers come bundled with basic software. This is usually a 'sawn-off' version of generally available software with some of its cleverer options removed, so that you will be encouraged to go out and buy the full edition.

Simple software often gives you two main options for burning a complete CD. These are Track At Once (TAO) and Disc At Once (DAO). Track At Once means literally this: each CD track is burnt as a separate action by the software, with the laser being turned off between tracks. This usually imposes 2-second pauses between tracks. Disc At Once burns the entire disc in one go without switching off the laser. With simpler software this provides no gaps at all between tracks. For more information, see Chapter 15.

3.14 DVD

Digital Versatile Disc is as much an audio format as its more common video usage. At the time of writing DVD is only beginning to show its full potential, although it promises to have most of the advantages of CD but with five to ten times the capacity. The standard also makes provision for more sophisticated audio formats, notably '5.1' channel surround sound. Audio CDs are restricted to two-channel stereo. Playback machines, which can also play audio CDs, are rapidly becoming as widespread as CD players.

3.15 MIDI

Many digital audio editors are also set up to handle MIDI. If you are just using audio, then these facilities can be ignored. MIDI is only useful if you are planning to include synthesized music sessions within your programme, or you are synchronizing the audio to other events, such as a *Son et Lumière*.

MIDI data records performance data, NOT audio. It is the modern electronic equivalent to the punched sheets that controlled player pianos and fairground organs.

MIDI can be used to synchronize different PC programs. For specialized purposes, like exhibitions and *Son et Lumière*, MIDI can be used for external control of the audio output. (See Appendices 2 and 3 for more information).

3.16 Control surface

Audio software often has lots of graphical displays of knobs and sliders for the user to operate. With a mouse, you only have one 'finger' to operate the controls. Today, external specialized control surfaces are becoming increasingly available at less and less cost, and these contain physical knobs and switches that operate the graphical ones on the screen. Some of these look like full-scale mixers, although no actual audio goes the surface (just as no audio goes through the mouse when you are controlling sound with it). These connect via USB, Firewire or occasionally MIDI, and send control signals to the program.

A second use of these is to get you physically away from the computer. If you are playing an electric guitar, its pickup can also pick up buzzes from the monitor. For this sort of use you don't actually need lots of faders, but just a few knobs and switches to control the recording software.

4

Recording

4.1 Responsibilities

All the equipment, bells, whistles and toys are of no use whatsoever unless there is something worthwhile to record – something your listeners will want to hear. Acquiring that material is the first task.

At the more complex level, you may be editing something that has been put together in a studio, or maybe a recording of a live event. In the context of this book, this means that the recording has already been done; you are presented with a stereo mix on analogue tape, a DAT or even a multitrack mix on an eight-track video cassette-based digital multitrack. These days it may even be on a removable hard disk.

If you have a suitable sound card, then you may be making the original recording using *Cool Edit*. In practice, most people, most of the time, are using something external to originate material. It may be off-air from radio, from a CD or any number of delivery formats (see section 5.5, Recording).

Your first responsibility is to ensure that you can handle the delivery format. An eight-track ADAT tape is fine only if you have an ADAT machine and your computer is set up to record its tracks, either by analogue or using the optical ADAT eight-track digital interface.

Even with DAT, you should have specified the sampling rate you require. This should normally be the rate you intend to use for the final edited version: 44.1 kHz for compact disc-related work, or maybe 48 kHz for material destined to be broadcast. Do not use lower sampling rates, or bit rates lower than 16, even if material is destined as a reduced data Internet sound file. You never know when there will be another use for the audio, and the better the input to a poor reproduction system, the better it will sound.

However, much of the material put together with a PC audio editor is likely to be less ambitious, and self-acquired; probably using a portable recorder, preferably digital, with an ability to transfer digitally to your computer.

A common way of obtaining your material is the interview; so here are a few guidelines to get the best out of it. These guidelines are as applicable to the promotional sales interview with the chairman of the company, as for a news interview for local or national radio.

4.2 Interviewing people

Preparation

Editing starts here, so consider the following so as not to waste your own or other people's time and resources:

1 What is the purpose of the interview? Are you interviewing the right person? (Do they have the information you want? Do they have a reasonable speaking voice?) Remember that the boss's deputy or assistant often has a better finger on the pulse of day-to-day problems, so it can be useful to interview both, if you can.
2 Prepare your subject area for questions, rather than a long list of specific questions.
3 Choose a suitable and convenient location. A senior manager's office is usually suitable, if it has a carpet and soft furnishings; but it is extremely important to get the person out from behind the desk. While you will tell interviewees that acoustics is the reason, it is equally vital to be aware that people speak differently across desks. Sitting beside them on a sofa will usually get a much more human response.

Types of interviews

- *Hard interviews* are performed to expose reasoning and to let listeners make up their own minds. The interviewee comes in cold, with no knowledge of the questions. Hard interviews are commonly used with politicians or those in the public eye.
- *Informational interviews* are for getting as much information as possible, so these are likely to be a more friendly, conversational style of interview. You may prepare your interviewee with a warm-up by outlining the areas of questioning.
- *Personality interviews* are intended to reveal the personality of the interviewee.
- *Emotional interviews* are probably the most difficult type of interview, requiring the greatest tact and diplomacy from the interviewer. Such interviews are of the 'How do you feel?' variety used when interviewees may be under great stress following a tragedy, (although try to avoid the actual question 'How do you feel . . .?').

Technique

Make sure that you actually know how to use the recorder! Try it out *before* the day of the interview, and test the recorder before you leave base. Test it when you arrive and are waiting for your interviewee. Test it when you take level. Always keep the test recordings.

Always record useful information, such as the date and who you are going to interview. This will help speed the identification of a recording when the label has fallen off and when it has been transferred to computer (some DAT and Minidisc machines will record a time of day and date code on the recording. This can be useful, provided you remembered to set the clock when you took the machine out of its box).

NEVER say anything rude about the person or company. 'I'm off to interview that prat Smith of that useless Bloggo company' might get you a laugh in the office, but loses its edge when you find yourself accidentally playing it back to Mr Smith after he has given you level. Machines that only offer headphone monitoring are a help here, but spill from headphones can be very audible.

If using a handheld microphone, then sit or stand close to the person being interviewed. Again, get the interviewee out from behind the desk! Side by side on a sofa usually works very well. The microphone should be 20–30 cm from each of you; if you have to compromise, favour the interviewee.

Make a short test recording 'for level'. The nominal task is to set the recording level so that it does not distort or is so low in level as to cause noise problems when amplified for use. For level, do not ask a question that is going to be asked during the interview. Traditionally, interviewers were supposed to ask what their interviewee had for breakfast. These days it is likely to produce either a monosyllabic 'nothing', or a long and involved account of some special muesli. No, this is the opportunity to record the interviewee's name and title. Get the interviewee to say it, and use this as a way of checking the level on the recorder. As well as giving you a factual check – is the interviewee Assistant Manager or Deputy Manager? – it gives you a definitive pronunciation of his or her name; is Mr Smyth pronounced Sm-ith or Sm-eye-th? In some styles of documentary this can be used for interviewees to introduce themselves rather than the presenter doing it.

It is good practice to leave the level set from your previous test recording, so that the level is roughly right if you run in to a major news story on your way to or from the interview and need to 'crash start' a recording.

Another function of the level test is check for unnoticed problems. An air-conditioning noise that you do not notice 'live' may be very obtrusive on playback. Listen for, and anticipate, external noises – e.g. children playing, dogs barking, telephones ringing, interruptions.

There can also be unexpected electrical interference. Faulty fluorescent tubes can cause interference, as can radio and computer equipment. Beware especially of mobile 'phones. As well as the obvious annoyance of them ringing during an interview, audio equipment is very prone to interference from the mobile's acoustically silent 'handshake' signals with the network. Switch off your own mobile and ask your interviewee to do the same.

Relax the interviewee, if necessary, by discussing areas of questioning.

Keep eye contact, and make appropriate silent responses such as nodding or smiling. Ask short, clear questions, one at a time. Remember, we want to hear the interviewee not the interviewer.

Listen carefully, and keep your questions relevant to what is being said. Ask open questions – who, what, where, why, how? BBC Radio Four's first *The World at One* presenter, William Hardcastle, maintained that he added to this list the non-alliterative 'Is it on the increase?'.

Pre-recorded interviews

When pre-recording an interview, record the background atmosphere for 15–30 seconds before and after the interview to use when editing. In my experience, leaving a 2-second gap between 'Right, we're recording' and the first question will give you the cleanest pause.

If you are using two microphones with a stereo recorder, then deliberately record the interview 'two-track' so that the interviewee's voice is on one track and the interviewer's is on the other. Keep to a convention so that questions are always on the same track – 'the presenter is always right' was the tongue-in-cheek convention used in BBC Radio. In the days of quarter-inch tape, this also had the advantage that the answers could be heard on an old half-track mono machine that only read the top (left-hand channel) of the tape.

After the interview, play back the last few seconds to ensure that the piece has recorded OK. Ask if the interviewee feels that anything was left out.

Vox pop interviews

In the trade, interviews with people randomly selected in the street are known as vox pops (from the Latin *vox populi*; voice of the people). They present their own unique problems.

Always test the equipment before leaving base. Follow a closed question (e.g. 'Do you think. . .?') with an open question (e.g. 'Why do you think that?) to get interviewees to justify their opinions. Prepare several forms of the same question to put to interviewees.

Choose a suitable location away from traffic, road works, etc. (unless, perhaps, the item is about the traffic or the chaos caused by the road works). Approach interviewees with a brief explanation of why you're conducting a survey of public opinion. Keep the recording level the same and adjust the microphone position for quieter or louder interviewees. Record background actuality at the beginning and end of the tape.

(A comprehensive guide to interviewing is contained in Robert McLeish's *Radio Production – A Manual for Broadcasters*, 4th edition, published by Focal Press, 1999.)

4.3 Documentary

Documentary material will also use 'actuality'. This is material recorded on location, of activity, that can be used both for illustration and for bridging passages of time. Record lots of it! It is too easy to be so closely focused on the interview material that the non-vocal material is forgotten. Do not be in the situation of the producer who recorded a feature about a mountain climbing expedition and forgot to record any sounds of people climbing! (In this particular case the situation was saved by a single 2-second sequence being looped, repeated and used in a stylized way.)

Actuality is different from sound effects. Actuality is the real thing, recorded at the actual location. Sound effects come from a library and are used to fake actuality. This is fine for drama, but has only limited use in factual items. Using an FX disc of a Rolls Royce driving off would be fine as an illustration of a car driven by rich people, but not ethical if cued as 'Joe Smith driving off in his Rolls Royce after my interview'.

Music can enhance an item, but beware of copyright and performance rights. While the sound of a busker playing in the background is unlikely to be a problem, using a complete performance of a piece of music usually is.

4.4 Oral history interviews

In oral history interviews you are interviewing someone, usually an elderly person, about memories of decades before. This is possibly the first and only time the interviewee has been interviewed, and you want the person to be relaxed and speak freely. You need to be relaxed too, as these interviews are likely to be long. Sitting side by side on a sofa can work, but some people are unhappy about sitting so close to a stranger. Here, I have used a tip from Charles Parker who, with the Radio Ballads, pioneered British documentary radio. He suggested actually sitting or kneeling at the interviewee's feet. This has the effect of lowering your perceived status and emphasizes that the interviewee's memories and views are important. With one knee on the floor, the other becomes a convenient place to rest the elbow of the arm holding the microphone. Acoustically it is good because the microphone naturally falls about 45 cm from both you and the interviewee, with both bodies providing sound absorption.

4.5 Interviews on the move

Interviews on the move are perhaps not so much interviews as two or more people reacting to an environment – for example, observing badgers at night, a historical house or a carnival. The object is to capture as much of the atmosphere as possible while not making it totally impossible to edit.

In some ways, an ideal medium is a four-track recording. Two tracks can be fed from a stereo microphone to record atmosphere, and the other two by two mono microphones, one for the presenter – possibly a personal 'tie-clip' mic – and the other (handheld) for whoever else is talking from moment to moment. An alternative might be two small portables – Minidisc or DAT – one recording atmosphere and the other recording dialogue.

The major disadvantage of such an arrangement is that the technology is in danger of taking over. The person holding the mics and carrying the recorders may be employed for his or her knowledge of the subject, rather than technical skill. Here, a separate recorder operated by the producer can help out with collecting actuality.

It is very possible to get a dramatic and memorable recording using a single handheld stereo mic, but it is vital that the mic is not much moved about with respect to the speakers, as otherwise their voices will go careering around the sound stage. In noisy environments the mic should be as close as is feasible to the voices in order to get as much separation as possible, so that what is being said can be understood. Plenty of actuality should be obtained for dubbing over the edited speech. It is easy to add sound, but it is difficult or impossible to take it away. A sensible discipline is also required so that the unrepeatable sounds of an event are not spoken over, and a commentary is given that is articulate and will not need editing.

When editing a feature about, say, an outdoor market, the buzz of its activity is vital. While some material may be recorded in shops away from the noise of the market, probably the best material, editorially, will be acquired in the crowd. A beat pause before each question (or response) from the presenter is all that is needed to form the basis of a crossfade on atmosphere to a different section. Where the edit causes an abrupt change of character of the background, then a short section of separately recorded actuality can be added to cover the change (see section 9.5, Chequerboarding).

4.6 Recorders

The original portable acquisition format was direct-cut Edison cylinders. Later, 78-rpm lacquer discs were used by reporters during the Second World War. They were heavy – more transportable than portable – and could not be used 'over the shoulder'. With the perfection of tape recording by the Germans during that war, the 1940s and 1950s saw the rapid appearance of more and more useful portable tape recorders.

Tape (like disc) had the advantage that the recording medium was the same as the editing medium, which was the same as the final user or transmission medium. Removable digital media have the potential to be the same. However, at the time of writing, recorders using this sort of technology tend to be bulky and heavy and there is no agreed standard. Recorders are becoming available that use small hard disks with sufficient capacity not to need to compress the audio. They are either removable or can download their recordings via a USB or Firewire port, and download times of 40 times real time are achievable. The only disadvantage is that the recorder has to be brought back to the computer at base, or it has to have removable media such as a hard disk that plugs in to a PCM-1A socket with a suitable port being provided on the base computer.

Small recorders using Minidisc or DAT are very attractive. Ordinary domestic recorders can produce superb quality at prices a fraction those of professional gear. Also becoming readily available are recorders using memory cards. A problem with these, as well as Minidisc, is that they usually use 'lossy' data compression techniques. This can cause problems if the edited recordings are further going to be data compressed, especially using different systems. This is a hazard that especially faces broadcasters. If this is important to you, then use DAT but beware of its fragility and dislike of humid conditions. Although cheap, the trade-off is that domestic recorders use fragile domestic plugs and sockets. This is not a problem if your environment allows you to take care of the machine, or is sufficiently well financed to regard them as disposable.

Alternatively, an option is to purchase a case designed to protect the recorder. These make the whole package larger, but they have professional XLR sockets and also a high-capacity battery for longer recording times. There is also the incidental (but not inconsequential) effect of making the reporter look more 'professional'. This can make the difference between being granted an interview, or rejection. There are professional versions of both DAT and Minidisc recorders. These have XLR sockets, and are generally more robust than their domestic brethren.

Whatever and wherever you are recording, you need to get the best out of your microphone, and a basic review of acoustics and microphones may be helpful.

4.7 Acoustics and microphones

Acoustics

When making a recording, realize that the microphone will pick up sounds other than the ones you really want. So what does a microphone hear?

1 The voice or instrument in front of it
2 Reflections from the wall

3 Noise from outside the room
4 Other sounds in the room, including voices and instruments
5 General noises, if outdoors.

Separation

Much of the skill in getting a decent recording is to arrange good separation between the wanted and the unwanted sounds. The classic ways of improving this are:

- Multi-mics
- Booths/separate studios
- Multitrack recording
- Working with the microphone closer
- Using directional microphones.

Multi-mics

There are many ways of reducing the spill of unwanted sound while retaining the wanted. It can seem that the answer to most problems is to have as many microphones as possible, but in fact the opposite is true. If you have six microphones faded up, each microphone will 'hear' one direct sound from the sound source it is set to pick up. However, it, *and the other five mics*, will also get the indirect sound from its own sound source *and* every other sound source.

In my example, you will be getting six lots of indirect sound for every single direct sound. As a result, multi-mic balances are very susceptible to picking up a great deal of studio acoustic.

With discussions, a lot of mics very close together on a single table can provide extra control and give some stereo positioning. Avoid the temptation to seat contributors to such discussions in a circle with comfy chairs, coffee tables and individual stand mics. The resulting acoustic quality will be excellent; but unfortunately the discussion is likely to be boring and stilted. This is because once you go beyond a critical distance of about 120 cm, people stop having conversations and start to make statements.

A studio designed for multi-mic music recording will have a relatively 'dead' acoustic, with reflections much reduced. The disadvantage of this is that the musicians and singers cannot hear themselves properly. In a very dead studio, a singer can go hoarse in minutes! You have to substitute for the lack of acoustic by feeding an artificial one to headphones worn by the artist. Needless to say, such a studio would be a disaster for a discussion.

Booths/separate studios

Booths/separate studios give good separation, but with the penalty of isolating the performers from one another.

Multitrack recording

Laying down one track at a time while listening to the rest on headphones gives good separation, but it can be difficult to realize a sense of the excitement of performance.

Working with the microphone closer

When you work with the microphone closer, the direct sound becomes louder and so you can turn down the fader and therefore apparently reduce the amount of indirect sound from that microphone. This has the disadvantage that the microphone is much more sensitive to movement by the artist and other unwanted noises can become a problem – guitar finger noise is clearly audible, breathing becomes exaggerated, and spittle and denture noises can become offensive.

It is a matter of opinion just how much breathing and action noise is desirable; eliminate these noises totally and it no longer sounds like human beings playing the music, but if there is too much it becomes irritating.

Using directional microphones

Microphones are not particularly directional, and are likely to have angles of pickup of 30–45° away from the front.

High directivity mics will give poor results with an undisciplined performer who moves about a lot. The main value of directional microphones is not the angles at which they are sensitive, but the angles where they are dead. It is more critical to position a microphone so that its dead angle rejects the sound from another source. Here, a few degrees of adjustment can make a lot of difference to the rejection and no audible difference to the sound you actually want.

Orchestral music

Classical and orchestral music tends to be performed in a room or hall with a decent acoustic with a relatively small number of microphones – typically a main stereo mic plus some supplementary mics to fill in the sound of soloists, etc. Not only will the musicians be happier – many non-electric instruments are hard to play if wearing headphones – but also the larger number of musicians involved often makes it impractical to find that number of *working* headphones, let alone connect them all up and ensure that they are all producing the right sound, at safe levels.

Microphones

There is no such thing as the perfect microphone; a single mic that will be best for all purposes. While expense can be a guide, there are circumstances in which a cheaper microphone might be better.

There is often a conflict between robustness and quality. As a rule, electrostatic microphones provide better quality. They have much lighter diaphragms that can respond quickly to the attack transients at the start of sounds, better frequency responses, and lower noise figures. However, they are not particularly robust. They can be very prone to blasting and popping, especially on speech, where a moving coil microphone can be a better choice. Electrostatic microphones are reliant on some form of power supply, which will either come from a battery within the mic or from a central power supply providing 'phantom volts' down the mic lead. These can fail at inconvenient times.

Quality

Yes, of course we want quality, but what do we mean by this? We want a mic that can collect the sound without distortion, popping, hissing or spluttering. With speech, there is a premium on intelligibility that is only loosely connected with how 'hi-fi' the mic is. An expensive mic that is superb on orchestral strings may be a popping and blasting disaster as a speech mic.

Directivity

Directional microphones are described by their directivity pattern. This can be thought of as tracing the path of a sound source round the microphone – the source being moved nearer or further away so that the mic always hears the same level. At the dead angle the source has to be very close to the mic, and at the front it can be further away.

Beware that sound picked up around the dead angles of a microphone can sound drainpipe-like and can have a different characteristic from that at the front.

As a rule, the more directional a microphone is, the more sensitive it is to wind noise and 'popping' on speech. When used out of doors, heavy wind shielding is essential. Film and television crews usually use a very directional 'gun' mic with a pickup angle of about 20° either side of centre. These are invariably used with a windshield – the large furry 'sausage' often seen pointing at public figures in news conferences.

Figure 4.1 Cardioid

The most common directivity mic is the *cardioid* (Figure 4.1), so called because of its heart-shaped directivity pattern. Cardioid microphones have a useful dead angle at the back. Placing this to reduce spill is more important than ensuring that it is pointing directly at the sound it is picking up.

A so-called *hypercardioid* mic is slightly more directional (not more heart-shaped!) (Figure 4.2). It has the disadvantage that it has a small lobe of sensitivity at the back and is dead at about 10° either side of the back of the microphone. However, when used for interviews using table stands the dead angle is usually about right for rejecting other people around the table.

Another common type of microphone is called *omnidirectional*, as it is sensitive equally in all directions (Figure 4.3). Omnidirectional microphones are particularly suitable for outdoor use, as they are much less sensitive to wind noise and blasting. They can be better for very close working, where directivity is not important, as they pop or blast less easily – an important consideration with speech and singing.

Barrier mics (PZMs) can produce good results with little visibility as a result of being attached to an existing object (Figure 4.4). This can be anything from the stage at an opera to a goal post on

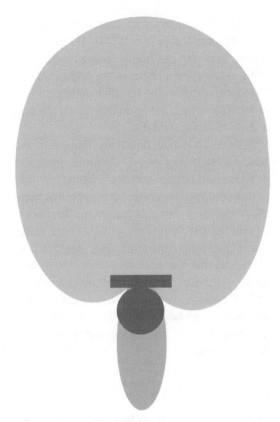

Figure 4.2 Hypercardioid

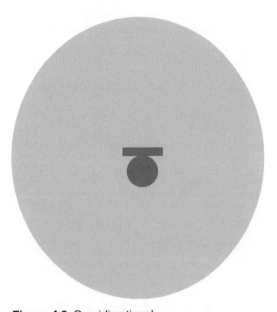

Figure 4.3 Omnidirectional

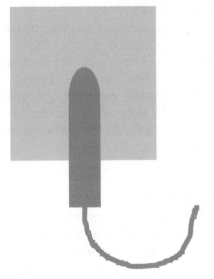

Figure 4.4 Barrier microphone

Figure 4.5 Hemispherical pick-up pattern

Figure 4.6 Figure-of-eight

a football field. They work best when attached to a large surface such as a wall, floor, table or baffle, and can be slung over stages using transparent sheet plastic as the baffle material to reduce visibility. The resulting pickup pattern is hemispherical; an omnidirectional pattern 'cut in half' (Figure 4.5).

A more specialist directivity pattern is the *figure-of*-eight (Figure 4.6). This microphone is dead top and bottom and at the sides, but 'live' front and back. This directivity pattern is most often found in variable-pattern microphones. Figure-of-eight mics can give the best directional separation provided that they can be placed so that the back lobe does not receive any spill, or can be used to pick up sound as well. Ribbon mics have the very best of dead angles and can be very useful when miking an audience that is also being fed sound from loudspeakers.

Variable pattern mics

These (usually very expensive) microphones offer nine or so patterns. They work by actually containing two cardioids back to back (Figure 4.7).

If only one cardioid is switched on then, as you would expect, the mic behaves as a cardioid. If both cardioids are switched on and their outputs are added together, then an omnidirectional response is obtained. If they are both switched on and their outputs subtracted, then a figure-of-eight pattern results.

The other intermediate patterns are obtained by varying the relative sensitivities of the two cardioids.

Beware; these derived patterns are never as good as the same pattern produced by a fixed pattern mic. The cardioids retain their physical characteristics, so an omni mic pattern derived from the two cardioids does not have the resistance to wind noise and blasting of a dedicated omni mic. The dead angle rejection or a derived figure-of-eight is not as good as a ribbon. Most microphones are 'end fire' – you point them at the sound you want. These mics are 'side-fire' with their sensitive angles coming out of the side of the mic.

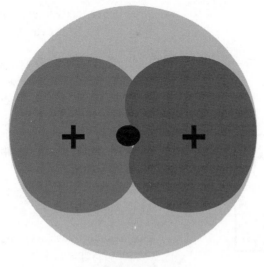

Two cardioids added
give omni response

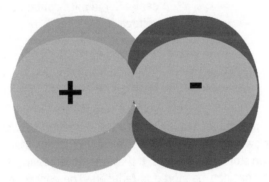

Two cardioids subtracted
give figure-of-eight response

Figure 4.7 Variable pattern microphone using back-to-back cardioids

Transducers

A microphone is a type of transducer – a device that converts one form of energy into another. In this case, acoustic energy is converted into electrical. Most microphones fall into one of two categories: electromagnetic or electrostatic.

Electromagnetic mics

Most electromagnetic microphones are moving coil. They are often called dynamic mics. They are like miniature loudspeakers in reverse; the acoustic energy moves a diaphragm attached to

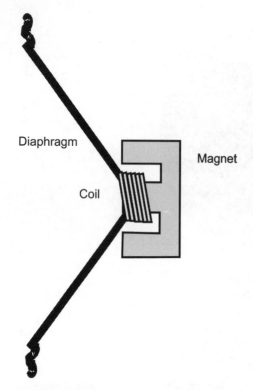

Diaphragm

Magnet

Coil

Figure 4.8 Moving coil electromagnetic microphone

which is a coil of wire surrounded by a magnet (Figure 4.8). This causes a small audio voltage to be generated.

A specialized form of electromagnetic mic is the 'ribbon' mic, which once was very widely used by the BBC. These have an inherent figure-of-eight directivity. They consist of a heavy magnet and a thin corrugated 'ribbon' of aluminium, which moves in the magnetic field to generate the audio voltage. They are 'side fire' mics.

Electrostatic mics

Electrostatic mics, also known as capacitor or condenser microphones, fall into two types. Many require a polarizing voltage of about 50 V for them to work. This is not a problem in a studio that can provide the necessary power as 'phantom' volts down the mic cable. Some mics contain 9-V batteries that operate a voltage boost circuit to provide their own polarizing voltage.

Electrostatic mics work by making the diaphragm part of a capacitor. A capacitor consists of two plates of metal placed very close together, and when a voltage is applied a current flows momentarily and the plate acts as a store for this charge. This is due to the attraction between positive and negative electricity across the narrow gap. If the plates are moved closer together, then this attraction increases and the plates can store more charge. If they are moved apart, then the attraction reduces and the amount of charge that can be stored is reduced. The consequence of the plate distance changing is a current through the circuit.

If one of the plates is the capsule of a microphone and the other is the diaphragm, made of a very thin film of conductive plastic, then we have the basis of a high quality microphone. The very lightness of the diaphragm gives it the ability to follow high frequencies much better than the relatively heavy assembly of a moving-coil microphone. As the diaphragm moves out, so the storage of the capacitor is reduced and charge is forced outwards. As the diaphragm moves inwards, the storage increases and the current flows in the opposite direction as the capacitor is 'filled up' by the polarizing voltage.

However, particularly for battery-operated mics, an 'electret' diaphragm can be used. This does not need the polarizing voltage, as a static charge is 'locked' into the diaphragm at manufacture. It is the electrostatic equivalent of a permanent magnet. While the technology is much improved, these tend to deteriorate with age.

There is a variant on the electrostatic mic where the capacitor, that is the diaphragm, is part of a radio frequency oscillator circuit. The changing capacitance changes the frequency of the oscillation, producing a frequency-modulated signal that is converted into audio within the microphone. RF capacitor mics have a good reputation for standing up to hostile environments, and are frequently seen as 'gun' mics used by film and TV sound recordists.

Other considerations

Robustness

Is the mic likely to be dropped, blown on, driven over? Does it blast on speech or singing?

Environment

Are you indoors or outdoors? Is the mic going to get wet or blown upon? Non-directional (omnidirectional) mics are least prone to wind noise, and this can be further improved by a decent windshield. In practice, using your own body to block the wind can be the most effective method of reducing wind noise.

Moving-coil (dynamic) microphones are the most resistant to problems caused by moisture or rain. Umbrellas are of little help in rain because of the noise of the rain falling on them! The windshield is usually enough to prevent rain from getting into the mic. If you ever need to record under water or in a shower, then a practical protection is a condom.

Size and weight

Presenters and singers do not like carrying heavy mics. Mics slung above people must be well secured by two fixings, each of which is capable of preventing the mic falling onto the audience or artists.

Visibility

This is affected by the size and weight of the microphone, and also its colour. Television producers of classical music programmes want minimum visibility. This may be helped by using small mics that are painted 'television grey'.

Operas on stage are often miked by six or so PZM/barrier mics laid on the stage and isolated from stage thumps by a layer of foam plastic (stage mice).

Power requirement

Total reliance on phantom power fed down the mic cable can mean losing everything if the power supply goes down. Batteries fail unpredictably. (Even if you thought you just put a new one in; in the stress of a session, it is easy to put the old one back in. Get into the habit of scratching a mark on the old battery before taking the new one out of the wrapper). Unpowered electromagnetic mics have definite advantages here.

Reliability, consistency and 'pedigree'

Microphones made by reputable manufacturers and bought through recognized dealers are likely to be more reliable and long lasting than an unknown make bought through a dodgy outlet. They are also likely to be consistent in quality, so one XYZ model mic sounds just like another model XYZ mic.

4.8 Multitrack

While most computers are equipped with single stereo sound cards, it is equally possible to fit cards that can handle more than two channels at a time. The main use for these is in music recording, but they can also be useful for speech, drama and documentary. However, multitrack is very intensive in its use of data. To state the obvious, a CDR with 80 minutes' capacity of stereo will only have room for 20 minutes of eight-track audio.

Many multichannel systems have quality benefits even if you rarely need multitrack recording. They use 'break-out' boxes so the critical audio components are outside the hostile environment of the computer. These boxes can be placed conveniently on the desk, while the computer can be floor standing.

If a long recording run is not required, then Minidisc, hard disk or even tape cassette-based multitrack recorders can be an option. There are also video cassette-based eight-track systems, which can provide about 40 minutes' continuous running.

While there may be a temptation to record a discussion multitrack with a view to pulling out extracts for a programme, in practice it is quicker (and cheaper) to have the discussion mixed straight down to stereo by an experienced sound balancer. You also save by not having to allocate time for mixing down the material from multitrack to mono or stereo.

5

Transfer

5.1 Reviewing material

Having acquired your material, how do you transfer it all onto your computer?

Most material will have to be copied, in real time, through the sound card of your computer, preferably using the digital inputs when the source is digital to preserve the best quality.

You should not waste this time, for it is now that you can review the material and make notes. You can save time where you know that things were redone, by skipping the false takes. This is especially easy if you have a note of ident points if you were using DAT or Minidisc.

Your notes should have a convention so that they mean something to you several days (or weeks) later. Write down who the speaker is and what tape or disc they are recorded on – you did label the recordings, didn't you?

Be consistent regarding how you log where things are in the recording. With reel-to-reel or compact cassette, you are usually stuck with arbitrary numbers produced by the playback machine's counter. These can be wildly different on a different make of machine.

Digital machines log progress as time, but there are different options. Quite a few portable machines actually record date and time of day continuously, and this does mean that on location you can use your wristwatch and a notepad to identify points. However, a lot of mains machines do not handle this information.

DAT machines usually have different counter modes:

- Counting time from when last reset
- Counting absolute time from the beginning of the first recording on the cassette
- Counting time starting from 00:00:00 each time a new track marker is encountered.

Minidisc recorders usually offer:

- Time elapsed of track
- Time left of track
- Time left of disc.

Once you have established which timing system you are going to use, make your notes. Unless you have good shorthand, you will not be able to transcribe what is said; instead, log each question with the time and then list the major points made in the answer. There are many formats possible, one such is:

Joe Smith 29th March

0'00	*?Why has this organization been set up?*
0'10	*Fulfil need by public*
0'20	*Contact point for victims*
0'35	*Money not available from government*
0'50	*?Should public money be provided?*
0'55	*Not a practical proposition*
1'05	*Get on with it*
1'20	*500 people already involved*
1'30	*?How long before organization effective?*
1'35	*First effects within 6 months*
1'45	*90 per cent after 2 years*
1'55	*Complete after 3 years*
2'10	*OUT ... 'Everyone needs this now'.*

When you are editing something that is scripted, then most of the edits will be overlaps (going back to the beginning of the sentence at the time) or retakes (sections redone after the main recording). It should not be necessary to edit fluffs, as they should be covered by overlaps and retakes. Pauses may need adjusting.

When marking an overlap, mark where it begins and how many times the speaker went back. There is nothing more irritating than editing an overlap and then discovering another one a few seconds later. Some people put brackets, the number of brackets indicating the number of repeats. You can play out at speed by holding down the mouse button on the spool button during playback. Identifying the repeats should be relatively easy, even at speed.

Better results are usually achieved by not cutting at the beginning of an overlap but instead a few words into it, because speakers tend to overemphasize the start because they are angry with themselves for making the mistake. With this in mind, when making notes mark where in the overlap you think will be a good return point.

It is amazing just how easily edits can be missed. The ear is sensitive to a break in the speech rhythm. It is always a mistake for speakers to apologize or to swear at themselves when they fluff. This has the uncanny ability to maintain the speech rhythm, and it is these edits that are most likely to be missed – especially in a news and current affairs situation. The professional way is to stop, give three beats pause, and then go back to the start of the sentence without comment. This will sound more natural if the edit is missed, but ironically that three-beat pause will usually mean that this does not happen.

5.2 Head alignment

If you are using an analogue recorder, a compact cassette machine or a reel-to-reel quarter-inch tape machine, then try to use the same machine that you used to record to dub your material onto your computer. This is because analogue tape systems are very sensitive to head alignment differences between machines. This gives a muffled sound. On a cassette, Dolby C doubly emphasizes this sensitivity. In practice, Dolby B noise reduction will reduce tape hiss to below the ambient noise in your recording environment without being hypercritical of head alignment.

Head alignment can be an issue with digital machines as well, and can mean that a recording will play on one machine but not on another. Try always to have the original machine available, just in case.

5.3 Digital

In some ways, digital recordings are the easiest to transfer. You do not have to worry about setting levels, as you did that on the original recording. If you got them wrong, then there is nothing that you can do to correct this until the digital audio data is on your computer. This is because the digital interface is merely transferring the numbers representing the audio that the original recorder laid down at the time of recording. Some sound cards allow you to alter the digital level, but this leaves you vulnerable to digital overloads. Correcting within the audio editor is much more flexible.

There are several standards for digitally interfacing your computer to a recorder, and the most common is known as S/PDIF (Sony/Philips Digital Interface). This comes in three main variants; one electrical and two optical.

Electrical

RCA phono plug

The electrical connection uses standard RCA phono plugs, as used for audio on a lot of domestic equipment (Figure 5.1). A single connection is used for both the left and the right sides of a stereo signal. Unfortunately many domestic digital recorders, while having an electrical S/PDIF input, do not have an equivalent output.

Mains operated machines will usually have an input and output in optical form, but many portable machines have no digital output at all. A cynic might believe that manufacturers think that their machines will only be used for copying CDs, rather than for creative work.

For best results you should use phono leads designed for digital connections. While any old audio leads you happen to have lying around may get you out of a jam, they can lead to unpredictable problems by introducing errors because they are not designed for the supersonic frequencies involved in digital transfer. Purpose-designed digital leads tend to be thicker than audio leads.

Figure 5.1 RCA Phono plug

Figure 5.2 TOSLINK optical plug

Professional set-ups will utilize balanced AES/EBU connections using XLR plugs. There are high-end sound cards that can use this format. However, S/PDIF inputs and outputs are usually also available.

Optical

Instead of electricity, optical connectors use modulated red light to transfer the data. This light is visible to the naked eye when a connector is carrying an output, and is extremely useful in avoiding confusion between input and output leads.

There are two physical standards. The original, usually found on mains operated equipment, is known as TOSLINK (from TOShiba Link; Figure 5.2). Small portable machines use optical connectors the same shape as audio 3.5-mm minijacks (Figure 5.3). Some machines have dual-function sockets, where the socket can be used either for audio or for optical digital. Both connectors use the same signal, which is S/PDIF in light form, so machines using different connectors can be joined using a minijack to TOSLINK lead.

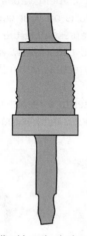

Figure 5.3 'Minijack' optical plug

Some early DAT machines had a proprietary connector at the machine end and a lead terminating with a TOSLINK plug. This can lead you to find yourself wanting to copy from recorder to recorder to 'clone' a tape, but being faced with joining two TOSLINK connectors. While not recommended, this can be done. Here the plastic tube that contains the ink in a cheap ballpoint pen can come to the rescue. Snipping an ink-free centimetre off the top will

Figure 5.4 Joining two TOSLINK plugs

provide you with a suitable 'gender-changer'. Make sure that you trim it short enough for the ends of the TOSLINK connectors to touch (Figure 5.4). This technique can also allow you to join two optical leads together. Yes, of course you should have bought a longer one, and you will tomorrow, but it can save the day today!

Optical leads are relatively fragile in that they will fracture if bent too far, as they are made of transparent plastic and not copper. They are surrounded by opaque plastic. Because the signal is conveyed by light, the connection is immune to interference from electrical equipment.

SCMS and copy protection

As well as the numbers representing the audio signal, the S/PDIF signal contains extra data, including various 'flags'. Normally these need not concern the user, except for the flags indicating copy protection. Literally, the system is specified so that a digital flag can be raised that will tell a well-behaved digital recorder input not to accept the data.

The all-or-nothing approach of a single flag was thought to be too crude, and the SCMS (Serial Copy Management System) was devised. This allows a single generation digital copy to be made. However, if an attempt is made digitally to copy that digital copy, then that is prohibited. This is an attempt to reduce multiple digital cloning of commercial recordings. Annoyingly, it is made to apply to your own recordings as well. This only becomes a problem when copying from an SCMS recorder to another SCMS recorder. Digital interfaces on computer sound cards ignore the system, and analogue copies always remain possible.

However, the original copy protect flag can be a problem when copying from a professional DAT machine to a domestic DAT. Professional machines use a variation on S/PDIF called AES/EBU, which is almost (but not quite) compatible when fed to a phono plug like an S/PDIF signal. The flags are different; a professional machine will often set the flag that means copy protect to a domestic machine, but not to itself. This leads to a situation where you can copy from the domestic machine to the professional machine, but not in the other direction!

Clocks

Stereo sound cards are relatively straightforward, but if you are using a multiple input sound card you will need to set the card to be controlled by the external data by switching it to the external incoming data clock (see Appendix 2).

USB/Firewire

There are digital audio recorders that record onto their own hard drive or a flash memory card. The one illustrated in Figure 5.5(a) can transfer via the USB port, giving all the advantages of digital transfer along with a speed around 40 times faster than real time. Effectively, for the time of connection the recorder becomes part of your computer, and you are copying a files from one storage device to another. With USB or Firewire you can connect or disconnect the recorder without having to reboot (Figure 5.5(b)).

SCSI/IDE

Some recording systems allow direct connection to a computer via a SCSI bus or a proprietary connection of their own. Alternatively, they may use removable hard disks. Contained in caddies, these can be unplugged from the recording device and slotted into the computer you use for editing. These often use the SCSI connection system, but IDE systems also exist. In the best implementations the inserted hard drive is allocated a drive letter by Windows, and can be edited directly by *Cool Edit* or any other editor. These disks can contain many gigabytes of data and hence many hours of work, so a backup copy should be a high priority. Carrying 20 hours of audio across

(a)

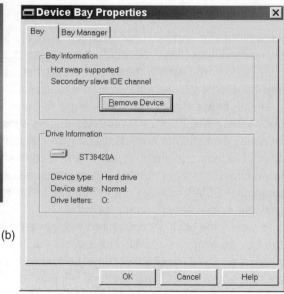

(b)

Figure 5.5 (a) Courier portable recorder from Sonifex Ltd. using a PCM1A hard disk; (b) Dialog to permit removal of caddy hard disk

town on a single disk is quite frightening, as it is vulnerable to so many possibilities of damage – from being dropped to being stolen. For physical protection, disks should be carried in a small, foam-filled flight case.

CDs

CDs can be played by the Windows CD player, or *Cool Edit Pro* has a CD player built into it. This is fine for listening to select material, but transfer is best done digitally. This is done either by using CD 'ripping' software, often supplied by the manufacturer of the CD-ROM drive, or by having your machine set up so that Windows 'sees' the audio tracks as wave files (see section 7.2, Blue bar blues). Version 2 of *Cool Edit Pro* has CD ripping built into it, but you may find that setting your CD-ROM to see CD audio tracks as wave files is much quicker. As well as having CD ripping built in, Version 2 can read audio CDs directly. Even so, the option of setting Windows to see CD tracks as wave files is extremely convenient, as then every audio program you use can read CDs.

When they were introduced, there was a lot of hype about CDs being indestructible. Most people have discovered that they are not, except in comparison with LPs. Unlike LPs, which will always produce something no matter how crackly, they share with other digital media the tendency to work perfectly or not at all. While inevitably the best advice is to look after your CDs and handle them by their edges, failed CDs can usually be recovered.

CDs play from the centre outwards. This is to allow easy compatibility between 3-inch and 5-inch discs. The beginning of the recording contains a 'table of contents'. While this is particularly heavily error corrected, it means that if the track is covered by a mark the CD will fail even to be recognized by the machine. If this happens, examine the playing surface for a mark, or look for marks at the inside of the playing surface.

CDs slow down as they track towards the outside, and this gives the digital track a constant linear velocity. The LP had a constant angular velocity, and the speed of the playing surface past the stylus decreased as it went towards the centre. This means that half way through the playing time of a full CD is not half way across the disc, but further out.

If the CD is jumping and skipping a section, this is likely to be due to a scratch or an opaque mark. If the CD repeatedly sticks at a section, this is probably caused by a transparent mark (like jam). This bends the laser and thus confuses the player as to where it is.

Armed with this information, it soon becomes possible to work out where to look and what to look for. Because of the nature of the digital track, scratches radiating from the centre are much less likely to cause problems compared with scratches running along the line of the track. This is why it is always strongly recommended that CDs be cleaned with radial movements rather than the instinctive rotary action (Figure 5.6).

The first emergency treatment is to wash the CD. Smear washing-up liquid over the surface and then run water over it from a tap, helping the washing-up liquid off with inside-to-out radial movements of your fingers. Cold or warm water is fine, but it is probably best to avoid hot water; if it is too hot for your fingers then it is

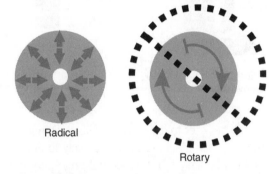

Radical

Rotary

Figure 5.6 Always clean CDs by wiping radially

too hot for the CD. Shake the remains of the water off the CD and look to see if there are any obvious marks remaining. These may need individual attention, again with the washing-up liquid. Try the CD in the player (any slight remains of water will be spun off in the player).

A really persistent mark or scratch should now be identifiable. If rescue has still failed, then the mark can be cleaned with brass cleaning wadding. Remember to rub radially.

It may be that this still doesn't work for you. If your CD player's laser lens is dirty, this will make it less able to correct errors. Clean the lens regularly with a special lens-cleaning CD, which has little brushes on it that clean the lens as the track is changed – usually between 1 and 3. Similar cleaners are also available for Minidisc machines.

If there is still a problem then try a different make of CD player, as different manufacturers have different error correction strategies. If you are 'ripping' the audio digitally straight off the CD, see if the CD-ROM has management software that will slow down the process. A CD that will not rip at ×12 speed may be OK at ×4, ×2 or ×1. If all fails, try the audio output of the CD-ROM drive. The 'View/show CD player' menu option switches on an inbuilt CD player for controlling the CD as an audio out. This can also be useful for auditioning material while you are doing other things. Taking the audio out of an external CD player can also be tried.

5.4 Analogue

Provided care has been taken to get the levels right on the original recording, analogue transfer can be entirely satisfactory. You should make sure that your system is connected properly and is well earthed. The playback should be checked for hum pick-up. This can be due to electrical equipment

Figure 5.7 Tip, ring and sleeve jack

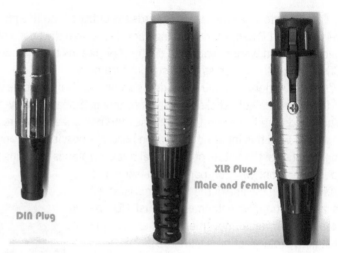

Figure 5.8 XLR connectors

on the desk, or to the computer itself. Running the dubbing lead near to a power supply (including the playback machine's own) is likely to introduce hum. Running the lead near to the computer's monitor is likely to introduce not only hum, but also buzzes and whistles.

Keep the dubbing lead as short as it practical, as very long runs will lead to loss of high frequencies unless you are using fully professional equipment and 'balanced' connections with tip, ring and sleeve jacks or XLR connectors (Figures 5.7, 5.8).

Normally you will connect the 'LINE OUT' socket of the playback machine to the 'LINE IN' socket. The most common connectors at each end are 3.5-mm minijacks (Figure 5.9); however, the better

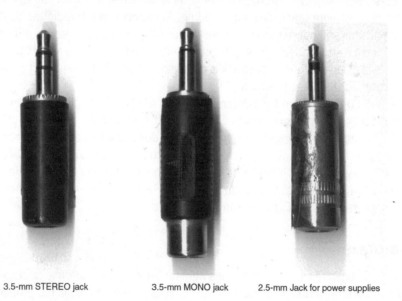

3.5-mm STEREO jack   3.5-mm MONO jack   2.5-mm Jack for power supplies

Figure 5.9 Stereo and mono 3.5-mm minijacks with 2.5-mm plug for comparison

quality sound cards have 'break-out' boxes. These either sit at the end of a thick flying lead, or are mounted in one of the disk drive compartments of the computer. They will have phono, quarter-inch jack or XLR connectors, often with a balanced connection option for use with professional machines.

The LINE OUTs of playback machines are of a fixed level, so level control has to be done on the computer. Using the headphone output instead is an option, and can give excellent quality. It can also produce crackles and distortion if the volume control is dirty!

Gramophone records

Despite the popularity of CDs, there are still occasions when old LPs are the only available source of a piece of music or even actuality. It makes sense, therefore, to get the best possible sound out of them before resorting to software solutions within the computer.

Gramophone turntables cannot just be plugged straight to a computer sound card. Not only do they have an output much lower that the line level required by the sound card, but they also need what is called RIAA equalization. This is a standardized fixed top boost and bass cut made on recording that has to be corrected at playback.

In the past, this was done by the 'phono' input of a hi-fi amp. Nowadays, many do not even possess such an input. Maybe you have an old hi-fi amp that can be pressed into service. The TAPE output can be fed to the computer's sound card.

However, such a complicated device is not needed. Cheap battery-operated amplifiers are available, and they deliver an equalized line level output suitable for your sound card. Being battery operated, they should have a lower hum level than conventional amplifiers. A typical example is illustrated in Figure 5.10.

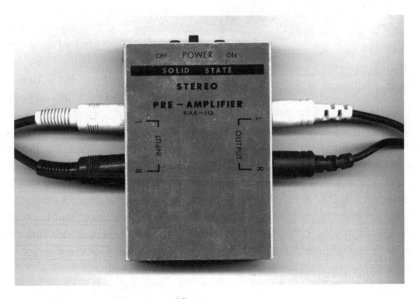

Figure 5.10 Typical record turntable preamplifier

Some can double as simple mic amplifiers, if required, by switching off the RIAA equalization.

It can be useful to bypass the RIAA top cut when transferring old recordings. The top cut softens clicks, and de-clicking software can work better without it having been applied. You then correct the recording afterwards.

If you are transferring 78 rpm recordings then RIAA should then definitely be left out, as they did not use this system. They did use some form of equalization, but there were more than a hundred different ones used by different companies over the half-century reign of the medium.

5.5 Recording

Before making a recording you have to tell the program how you want to record. Some audio programs, like *Pro Tools*, will also ask you where you want to record before you start. *Cool Edit* uses buffer files, and you can decide on file names and paths after the recording. You can set where these buffer files are by changing the temporary directories to be found in the system tab of the settings dialog. There is a speed benefit by having them on physically separate hard drives. This means that

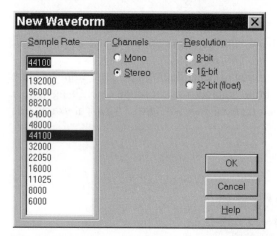

Figure 5.11 Setting recording options

copying is achieved by two heads moving across their separate drives rather than one head clacking back and forth between different parts of the same disk.

Click on 'New', and you will see a dialog asking you to set various options (Figure 5.11). This will be based on what you set previously, so for most people this becomes a simple matter of clicking 'OK'.

The sampling rate is how many numbers per second are used to define your audio. For high quality use, there are two main standards. CDs use 44 100, and if your material is going to end up on CD audio then this is the rate to choose (CD-ROM data can be any sampling frequency). The other commonly used sampling rate is 48 000, which is required by some broadcasting companies that accept data files. If your material is sourced from (or is going to end up on) DAT, then this is the frequency to use. Many non-pro DAT machines will only record analogue signals at 48 000 (or 32 000 at their Long Play speed). If you are making a digital transfer, it is vital that the sampling rate set matches the rate used on the recording. The sampling rate of a recording can be converted within *Cool Edit*, but this is a relatively time-consuming process.

The mono/stereo option is straightforward. Operationally it is much simpler to record everything stereo, even mono sources. A stereo transfer of a mono recording can help identify problems such as head alignment. Sometimes the left and right channels will be of different quality, and you can choose the better one.

However, mono files are half the length of stereo files, and if everything is in mono and of good quality then make the saving. Beware that CD audio has no mono mode, so burning to CD may need stereo files. However, many CD writing programs have the ability to convert mono to stereo on-the-fly, thus removing this constraint on the use of mono files. Minidisc does have a mono mode, and a 74-minute Minidisc can record 148 minutes of mono, an 80-minute disc 160 minutes.

The resolution sets how big the numbers that represent your audio can be, and thus the number of separate discrete levels that can be represented. This directly affects the noise level of your recording. Eight-bit only has 256 levels and is, as a result, very hissy. It should be regarded only as an end-user format. With suitable control and processing, 8-bit files can sound surprisingly good. However, that processing has to be done at a higher resolution.

Sixteen-bit files are the standard, match CD and DAT, and will be your normal choice. Thirty-two-bit files give the highest quality and, while taking up twice as much data space, give the highest quality results when being modified. If your acquisition source has higher than 16-bit resolution, or you are dubbing from an extremely high quality analogue source (e.g. 15/30 ips Dolby reel-to-reel tapes), then, provided that your sound card has better than 16-bit analogue to digital conversion, 32-bit will give extra quality. Similarly, if it can cope, doubling the sampling rate to 88 200 or 96 000 will further improve the quality; however, your resulting wave files will be enormous.

The disadvantage of working at 16 bits is that this is what the domestic consumer has at home. A 16-bit recording will have to have some headroom left in case of unexpected loud peaks, and quiet passages may have to be amplified. This can mean that your final recording is, in parts, only equivalent to a 14-bit or less recording. Working at 32 bits allows you to correct levels, normalize, and then reduce to a full 16-bit master.

Setting the level

Oddly, programs for audio rarely have input gain controls. The job is usually done either by a utility provided by the sound card, or by the Windows sound mixer. This is accessed by double-clicking on the yellow loudspeaker icon on the task bar (single-clicking produces the Windows loudspeaker volume control). This will open the playback mixer. To get at the record controls, you need to click the properties item in the options menu and then click the record option (Figures 5.12–5.14). You can then select which source you wish to record from. If you have a specialist card, maybe with several separate inputs, then it is likely to have its own manager for routing the audio. Here you will have to tell the editor which card to use.

In *Cool Edit*, this is done using the Options/settings menu (F4) and clicking on the 'Devices' tab. This will provide information on the card's capabilities and selection of the record and playback options.

Multi-channel sound cards usually present themselves to Windows as a number of separate stereo sound cards – so a card with eight

Figure 5.12 Audio mixer properties

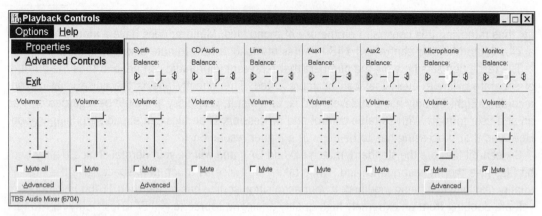

Figure 5.13 Audio mixer – playback

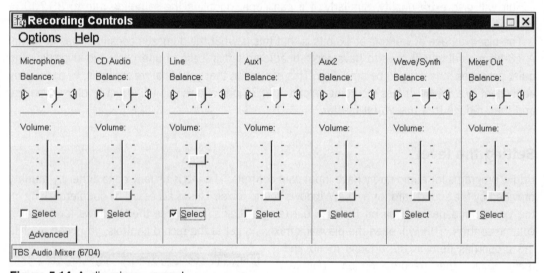

Figure 5.14 Audio mixer – record

analogue inputs and outputs plus a stereo digital in/out will appear as five separate cards. This is for the single wave view mode. In multitrack mode, each track can be separately selected to a sound card input or output (Figure 5.15; see Chapter 9).

Because running level meters requires a fair amount of processing time, most editors leave them off except when actually recording (playback is less of a problem as the system already 'knows' what the levels are). It is this processing overhead that makes programmers avoid emulations of analogue meter pointers, like VU meters and peak programme meters.

Options/monitor record levels (F10) selects the meters in *Cool Edit*. The program counts this as a pseudo-record mode and the red record light comes on, so you can only go into record via stop. However, this visual preview gives high resolution meter monitoring. When you actually go into record you do get a meter display, but this is less precise than the preview mode; this is more than compensated for by the graphical display of your recording building on your screen.

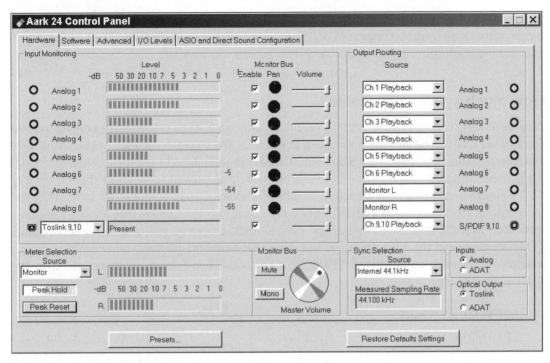

Figure 5.15 Example of a multitrack sound card control panel

Recording a file

Switch off the meters by clicking the stop button or using F10. Clicking the record button will switch off the meters as well, but you will have to click it again actually to go into record. The meters will now reappear (but be less responsive than before). The waveform of the recorded audio will be drawn on the screen in blocks. Just as the waveform is getting towards the right of the screen there will be a redraw to move the window along to keep up with the pointer, which will remain static (this can be switched off if you wish).

You can 'crash' start into record from a blank window by clicking record and pressing ENTER to the dialog requesting the record parameters. This will utilize the settings that you last used.

The recording is stopped by clicking the stop button, at which point the window will redraw to show you the entire waveform.

At the bottom of the screen is a list of your disk resources (Figure 5.16). The first box shows the sampling rate, bit rate, mono or stereo; the next box shows how many megabytes your recording

Figure 5.16 *Cool Edit Pro* disk usage display

has used, followed by how many hours, minutes, seconds and frames. These two increment in blocks, rather than by the second, as *Cool Edit Pro* allocates the space this way. The next two boxes give you the equivalent figures, but for how much space you have left.

How much space is available?

This raises a vital issue. What is actually meant by how much space is left? Some editors, like *Pro Tools*, insist that you define where you want to record at the beginning of each session. *Cool Edit Pro* records to its own buffer. This is different from where you intend to put the finished recording. An ideal set-up system will put *Cool Edit Pro's* two buffers on separate disk drives (both preferably SCSI). Finished recordings can be saved to different folders or partitions on the same disks (or, indeed, to other drives on the system). Personally I have two SCSI drives that are used as audio work files, with IDE drives used for storage and non-audio programs. I also set Windows to put its TEMP file on one of the SCSI drives, as this gives speed benefits for every other program on the machine. To change the Windows TEMP file destination, two lines have to be added to the AUTOEXEC.BAT file:

```
SET TEMP=G:\TEMP
SET TMP=G:\TEMP
```

(where 'G' should be replaced by the letter of the drive you want to use).

Topping and tailing

Once the recording has been finished, top and tail it by removing any extra material at the front and end. Ends usually sound cleaner if you use a fade out transform on the last half-second or so. This guarantees that your recording ends on absolute digital silence, and removes a potential source of clicks.

Mobile telephones

Mobile 'phones have become almost universal. They are invaluable, especially in a business where gigs can be set up at short notice. However, you should be aware that they can cause severe interference with audio gear. Semi-pro equipment is particularly vulnerable as it uses unbalanced circuits, but even the most expensive professional products can be affected. This why they are banned in hospitals, where they can interfere with medical equipment. When you are transferring your audio to the computer, you should avoid having your mobile switched on nearby.

Each mobile has a radio transmitter, and it is this that causes the interference. The interference shows up as burbling noises, or even as semi-musical arpeggios. Therefore, it is best practice to turn off mobiles while in the same room as audio equipment in use. It is particularly impolite not to turn off your mobile if the audio gear is being used by someone else! Interference can be caused even in standby mode, although this will come in short bursts spaced minutes apart.

Essentially, a mobile 'phone has four modes:

1 *Off*. This is the only safe mode.
2 *Ringing*. Although the ringing is incoming, the phone is transmitting data to the network to tell it where it is and that it is responding to the call.
3 *Talking*. The transmitter is sending your side of the conversation to the network.
4 *Standby*. Here the transmitter is usually off; but every now and then it wakes up and sends a 'Hi there!' message to tell the network that it is still about. These messages can be even more frequent if the phone is out of range of the network, as it will be trying to re-establish contact.

6

Editing

6.1 Introduction

Cool Edit Pro is two editors in one; linear and non-linear. You simply toggle between them by clicking the icon at the top left-hand corner of the screen. It changes in appearance depending which mode it is in.

When it shows a stylized view of three tracks you are in the linear single wave view mode, and clicking it will take you to the non-linear multitrack mode. It will then change to show a representation of a single wave, indicating that clicking it will return you to the linear mode (see Figures 6.1 and 6.2).

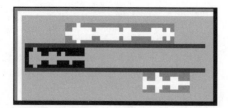

Figure 6.1 *Cool Edit Pro* non-linear editor icon

Figure 6.2 *Cool Edit Pro* linear editor icon

We'll look first at the linear editor, which physically changes the audio files so that data are changed when the files are saved.

When you first run *Cool Edit Pro*, you see a blank centre screen surrounded by a large number of icons (Figure 6.3). This should not be a cause for panic! The icons at the top can be ignored until you are familiar enough with the editor to find them useful. The most important controls are the transport and zoom icons at the bottom, and we will come to them as soon as we have loaded a file.

6.2 Loading a file

OK, maybe you have not recorded anything yet, but you will find that your computer already has a number of wave files. They have the file extension WAV (if you have set Windows to show

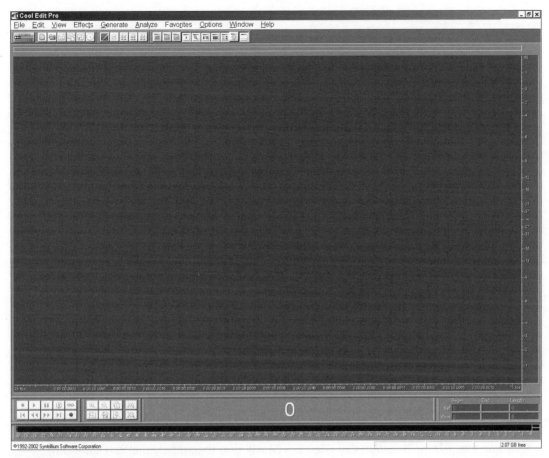

Figure 6.3 Entering *Cool Edit Pro*

extensions). You will find a lot of them in your C:\WINDOWS\MEDIA directory. These are the sounds that Windows can make when it is asked to undertake various actions. You may have turned some (or all) of these off, but the wave files are still there. Indeed you may decide that your first exercise with the audio editor is to make some new sounds to personalize your computer. Loading JUNGLEWE.WAV gives the display illustrated in Figure 6.4, which is a clap of thunder. On some versions of Windows it is called JUNGLE WINDOWS EXIT.WAV. The length is very slightly different owing to the non-audio data being changed. There are probably as many different installations of Windows as there are machines running it, and while the file I have chosen is available on Windows 95, 98 and ME CDs, it is possible that it was not copied to your hard disk when it was installed. This depends on what sound scheme options were selected. It is part of the Jungle Sound scheme.

In Figure 6.4, the status line at the bottom of the screen reveals that the recording uses 324 k of disk space, is 3 seconds and 19 frames long, stereo, at 22 050 samples per second (CEPs display in the illustration is set to the European TV standard of 25 frames per second. Don't confuse this with the compact disc digital structure, which has 75 frames per second).

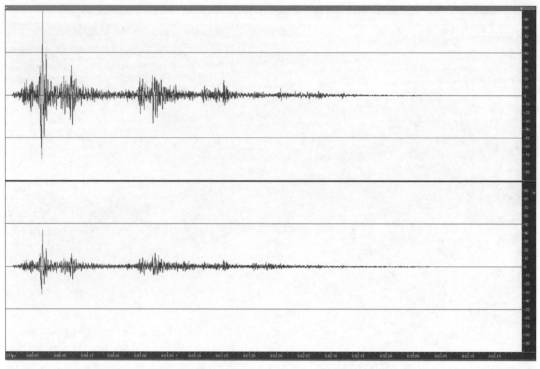

Figure 6.4 Display when JUNGLE WINDOWS EXIT.WAV is loaded

This particular recording looks as if it is low level, but in fact it is only 0.11 dB short of full modulation. How do I know this? Because I can get *Cool Edit* to tell me!

Selecting Effects/Amplitude/Amplify (Figure 6.5) will give the amplify dialog (Figure 6.6).

You will see the DC Bias and Normalization options in the bottom left-hand corner of the display. Clicking the calculate button shows the amount of boost required for full normalization.

In practice, most of the time you do not need all the options of the amplify dialog and can use the Normalize option (Figure 6.7).

This allows you to set the normalization level, choose whether to apply DC bias correction and, for stereo files, choose whether you want the left and right channels separately normalized or to keep the level change the same for both (this preserves sound images' positions in the sound stage).

Close examination of some waveforms can reveal that they have what is called a DC offset. I have simulated a 10 per cent offset in Figure 6.8. If you look at the detail, you can see that the whole waveform is offset slightly below the centre line. Offsets like this are highly undesir-

Figure 6.5 Select Effects/Amplitude/Amplify

```
Effects
  Refresh Effects List

  Invert
  Reverse
  Silence

  Amplitude        ▶   Amplify...
  Delay Effects    ▶   Channel Mixer...
  DirectX          ▶   Dynamics Processing...
  Filters          ▶   Envelope...
  Noise Reduction  ▶   Hard Limiting...
  Special          ▶   Normalize...
  Time/Pitch       ▶   Pan/Expand...
                       Stereo Field Rotate...
```

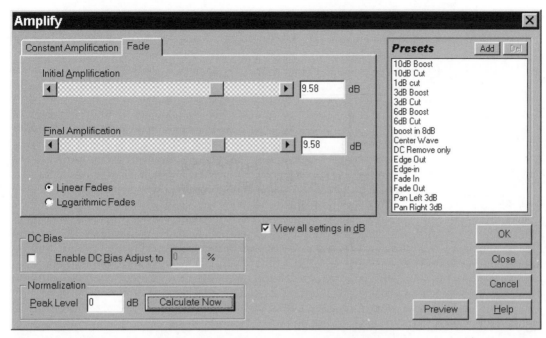

Figure 6.6 Selecting Effects/Amplitude/Amplify gives the amplify dialog

able, as they will produce a click every time the track is started or stopped and can cause clicks on edits.

A good quality sound card should not introduce an offset, but a lot of the cheaper ones do. This is not a disaster, as the offset can be removed by the editor. In *Cool Edit Pro* this can be done as part of the normalization process, when the level is corrected to produce full level.

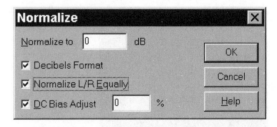

Figure 6.7 The normalize option

It has also an option to remove DC on-the-fly during recording. Beware that if you leave this switched on 'just in case', this will prevent you from doing bit-for-bit digital clones via *Cool Edit*.

If you actually want a DC offset, you can change the value from 0 per cent to the value you require. This may seem an unlikely need, but could be useful to correct an offset manually. *Cool Edit Pro*'s DC correction introduces a bass cut. If your recording has a deliberately extended bass response, you would not want to lose this (however, a sound card poor enough to have a DC offset is unlikely to have an extended response).

The Decibels format option sets the values entered into the 'Normalize to' box to show decibels rather than percentages.

The reason the audio looks so low level is that there is a brief moment at the start of the thunder where the level virtually reaches the peak. By zooming in on the detail, we can see that it is literally one cycle of audio from negative to positive. Because the stereo sound is not coming from the centre, the right-hand channel (bottom display) is even lower in level (Figure 6.9).

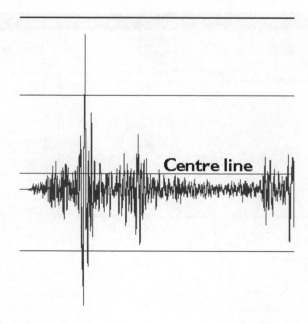

Figure 6.8 DC offset

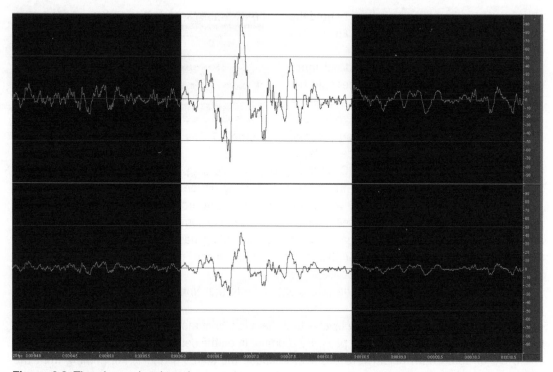

Figure 6.9 Thunder peak selected

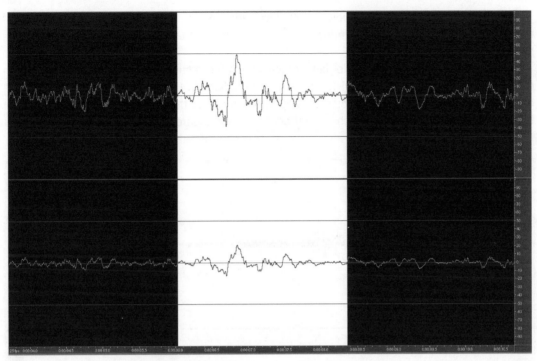

Figure 6.10 Thunder peak reduced by 6 dB

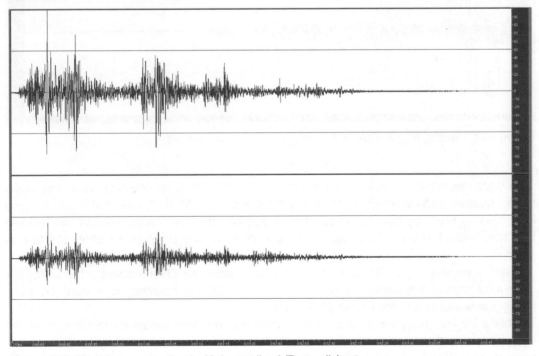

Figure 6.11 Thunder, renormalized with 'normalize L/R equally' set

Short peaks are a common problem with level adjustment. It is one of the prime skills of mastering engineers in the record industry to modify these so that the resulting CD can be made louder.

In this case, we can make a louder file by highlighting the section, reducing it by 6 dB and then renormalizing. We can also, if we wish, untick the Normalize L/R Equally box, to bring up the right-hand channel. Using the Edit/Zero crossings options can ensure that our selection does not stop or start in the middle of a cycle, which can introduce a click (Figures 6.10 to 6.12).

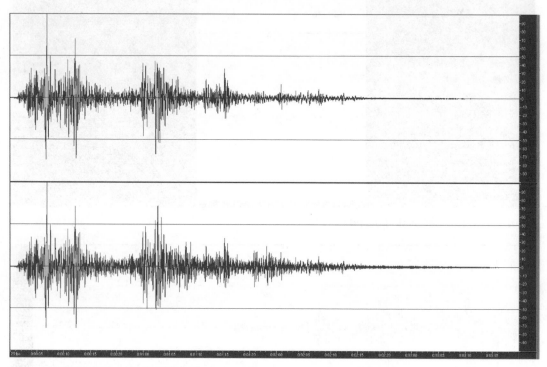

Figure 6.12 Thunder, renormalized with 'normalize L/R equally' not set

By doing this manipulation we achieve a louder recording, although the peak level remains the same. However, we have brought up the background noise as well, which may or may not be a good idea. You may like to try this, if you have this file on your hard disk. The process has now revealed two more peaks that are holding down the level. When do you stop? That's where skilled ears come in.

A similar process is gone through when music is mastered for CD. Removing a few peaks can make a commercial disc sound louder. This has become rather competitive, as no company wants its discs to sound quieter than those of its rivals.

While a full level recording is ideal for production use, this may not always be the case for the version heard by the end user. Windows sound effects are a case in point; it is generally desirable to set these 6 dB to 12 dB below the maximum recording level. This allows the PC speakers to be set at a good level for when they are reproducing music from a CD – or even your editing session, without the Windows sound effects blasting you out of the chair!

To modify recordings, we need to be able to navigate round a sound file. *Cool Edit Pro* presents the whole of the file when loaded. Personally I find this reassuring, and I have never got on with other editors that only show the beginning of a loaded file.

The keys to navigation are the transport and zoom icons at the bottom left of the screen (Figure 6.13).

The left-hand block is for the transport, and is reassuringly like a cassette recorder with a few extra options. Stop, Play and Pause are followed by the triangular play icon in a circle.

Figure 6.13 Transport icons

The last icon has an infinity sign. Clicking the standard play triangle symbol plays the file, as you would expect. However, it will stop when the playback cursor reaches the end of the screen – or the selection, if one has been made. In our present case this will be the end of the file. If, however, we were zoomed in, it would not be.

The triangle in a circle play button will also start the playback, but this will continue to the end of the file. The last icon sets up a loop mode so that playback never ends but loops round, either to the start of the file or to the start of the selection.

The second row of icons behaves like those on a CD player. The double arrow buttons 'spool' the cursor along the file (audibly if in the play mode). The other spool buttons take you to the beginning or end of the file, or to the next marker if you are using them. The final button, with the 'red LED', is the record button.

The playback 'head' is represented by a vertical line. This can be moved to where you want, merely by single-clicking the point with the mouse. If you are already playing the file, then a cursor line will appear where you clicked but playback will not be interrupted. Clicking the play triangle will instantly move the playback to the newly selected point. The space bar toggles playback mode on and off. If you are using this, then it toggles through stop before going to the new location on the next press.

You can select an area by dragging the mouse over the area you want. The alternative is to click one end with the left mouse button, and then click the other end with the right mouse button. The selected area can be modified with right mouse clicks. The selected edge nearest the mouse click is the one modified. By default, version 2 of *Cool Edit Pro* requires you to press control to change the selection with a right click, otherwise it gives you a pop-up menu. This can be reversed by clicking the Extend Selection option within the Edit View Right Clicks section in the General tab of the Option/Settings dialog. This reversed operation was how version 1 worked, and is, in my view, preferable.

You can also select the edit area 'on-the-fly' as you play the file. There are two keyboard short cuts, called 'Selection anchor left when playing' and 'Selection anchor right when playing'. 'Out of the box', *Cool Edit* sets these to the '[' and ']' keys. This means that if you press the '[' key while you are playing a file, the start of the selection will be set to the point playing at that moment. At the same time, the point where *Cool Edit* starts playing from when play is pressed is moved there as well. Pressing the ']' key marks the end of the selection. Pressing 'Delete' will then delete that selection while Play is continuing. Pressing the spacebar twice will stop and then restart play from your edit. You can then 'spool' back and check from before your edit. The spool controls are not allocated a keyboard shortcut by default, but this can be done (see section 17.3, Keyboard shortcuts). Being able to use the keyboard rather than the mouse can vastly improve the efficiency of a session where a great deal of edits are made. This is more typical of a speech editing session than a music one. This also transforms the use of the program for blind users, who can now edit by ear rather than being frustrated by the visual nature of PC audio editing.

The right-hand group of three icons controls zooming. The top three are absolute controls: '+' zooms in and '–' zooms out. This only affects the horizontal magnification. If you have a wheel mouse, then the wheel will also control zooming. The final icon restores the view to the whole of the file.

The bottom three icons of this block deal with the selected area of the file. The first fills the window with the selected area. The second icon zooms in on the left edge of the selected area and the third, correspondingly, zooms in on the right-hand edge of the selected area. This is particularly useful for fine trimming the start and finish of edit points.

There are two more zoom icons to the right of this group of icons. These '+' and '–' buttons control the vertical magnification of the audio. This is not needed very often, but can be useful to examine the character of low-level sounds such as noise.

It is very easy to examine a section of audio quickly. Click on the point you are interested in, and define a selected area by dragging or using the right mouse button and clicking on the 'Zoom to selection' icon. You can repeat this as many times as you like to get into really fine detail. You can even set the magnification to be so large that you can see the individual samples (Figure 6.14).

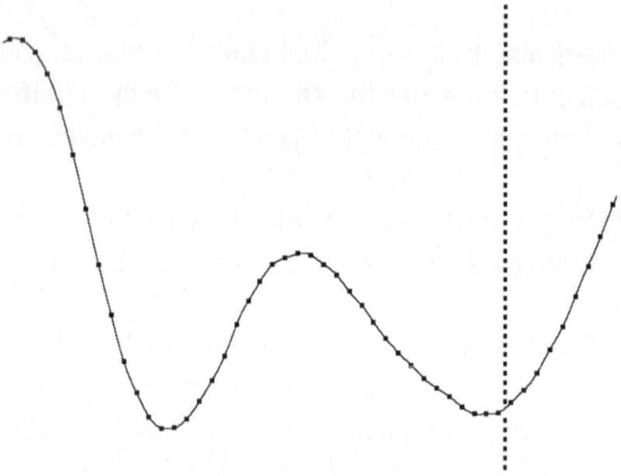

Figure 6.14 Waveform zoomed-in to show samples

These are represented as dots, and are joined by a line generated by the computer. This line is not simply dot-to-dot for cosmetic reasons, but will show what will happen to the audio after the output filter.

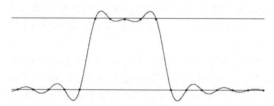

Figure 6.15 Five sample square pulse drawn onto waveform showing computer's interpretation of the result

If necessary you can move the dots, literally to redraw the waveform. This can sometimes be the only way to eliminate a very recalcitrant click that has slipped through declicking software.

In Figure 6.15 I have drawn a five sample click this way, to illustrate that the line joining the dots is more intelligent than it first appears; the wavy lines show what is known by audio engineers as 'ringing'.

6.3 Making an edit

Editing on a PC is best done visually. Some editors have a 'scrub' editing facility where the sound can be rocked backwards and forwards as on a reel-to-reel tape machine. People used to quarter-inch tape ask for this, probably out of fear or sentiment, but in practice nearly everybody edits visually. The technique is very simple, and the eye and the ear work together easily. It is much easier to learn from scratch than is scrub editing. Very soon you become able to read the waveform. For example you may know a particular point is overlapped three times, and it is often possible to spot the individual retaken sections and select the last one without listening to the intervening audio. If, on performing the edit, you find that you are wrong, then UNDO quickly restores the cut. By making 'bets' like this you can considerably speed up the editing process, particularly with scripted speech editing.

Editing audio is surprisingly similar to editing text in a word processor. If you use a word processor, then you already know the basics of editing audio. You use your mouse to select a start and finish point. The space between is highlighted by reversing the colours, and pressing DELETE removes the highlighted section. Alternatively, clicking Edit/Trim (or Control/T) will eliminate everything except the selected area (ideal for 'topping and tailing').

As already described, the selected section can be created in many ways. The most obvious is to place the mouse at the start of the section you wish to cut and drag the pointer to the end of the cut section. *Cool Edit Pro* also allows you to select the other end of the selection from the cursor by right-clicking on the point (while pressing control, unless you have reversed the action of the right-click). This works forwards or backwards.

6.4 Visual editing

This is all very well, but how do you find a phrase within an item? Here you use the zoom facility already described.

Figure 6.16 shows a short section of speech – 'The um start of the sentence'. The text has been added to the picture, and is not a feature of current audio editors!

How do we go about removing the 'um'? Clicking the play button with the mouse (or pressing the space bar) will play the section. A vertical line cursor will traverse the screen, showing where the playback is coming from. Your eye can identify the section that corresponds to 'um'.

Using the mouse, click at the start of this section. The cursor will move to this point. Press play again and the playback will start from where you clicked. You can hear if you have selected the right place, as if you have not there will be something before the 'um' or it will be clipped. Reclicking to correct is very fast.

Now identify, not where the 'um' ends, but where the next word starts – the 's' of 'start'. This time click with the RIGHT mouse button. This sets the end of the section to cut (Figure 6.17).

Hit play again and just the highlighted section will be played. Is the end of the 'um' clipped? Can you hear the start of the 's'? If not, then the edit is probably OK.

Now press the delete key. The highlighted section is removed (Figure 6.18).

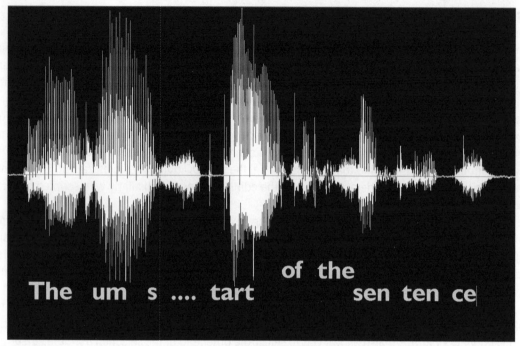

Figure 6.16

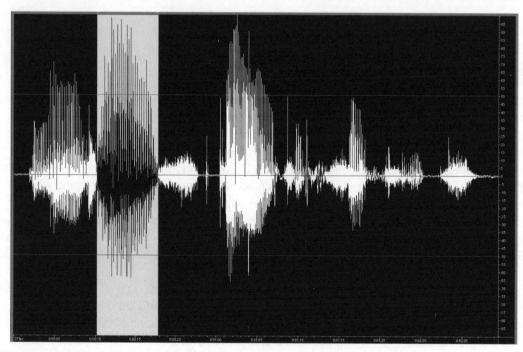

Figure 6.17 'Um' selected for removal

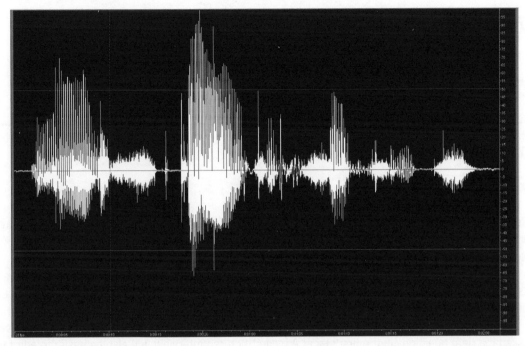

Figure 6.18 'Um' removed

Click a little before the join and play the edit. Is it OK? If so, go on to the next edit. If it is not, then click UNDO and you will be restored to the previously highlighted set-up.

You can adjust the edges of the selected area, by right-clicking near them.

At first you may well have to cycle round several times, but soon you will be getting edits right first time. Remember, even redoing the edit three or four times is still likely to be quicker than it used to be with using a razor blade.

7

Quarrying material

7.1 Introduction

Returning to base with material of exactly the right duration and needing no editing is a rare occurrence. If your recording is of a continuous event, then simply cutting it to time is all that is required.

With unscripted speech-based items you can return with substantially more than you will ever use, especially if you have recorded a series of interviews that you plan to 'quarry' out as extracts rather than using them complete.

However, it cannot be stressed too much that the very best way to edit an interview is to have asked the right questions in the first place! A well-structured interview will also be easier to edit for duration. However, circumstances can force returning with far too much material, especially if vox pop (in the street) interviews have been recorded.

Once upon a time there was no alternative; quarter-inch tape had been used and the only way to edit was with a razor blade, wax pencil and sticky tape. Copy-editing added extra time and was avoided.

7.2 Blue bar blues

No matter how fast your computer and its hard drives, audio files consume a vast amount of data compared with a word processor, and take time to be saved. Most editors have a very useful UNDO facility. This works by saving a copy of the section of the file you are changing onto hard disk.

All this takes time and slows down the editing process. This means that you should structure your material so as to avoid long audio files of more than 3–5 minutes. Put plainly, a single edit within a 30-minute file takes much longer to do because of the time taken to save that long file. A 5-minute file takes one-sixth as long to save. A strategy of one or two files per item within a longer piece will speed things along nicely.

Extracting sections of a file to new separate files is most easily achieved by highlighting the section and using the 'File/Save Selection' menu option.

Where you just want to copy and paste then use the clipboard, where, as with all modern operating systems and programs, Control/C will save the selected area. Control/X will cut it out and Control/V will paste it to the current cursor position. *Cool Edit*, as well as using the Windows clipboard, also provides you with five clipboards of its own.

However, there will be things you will want to do to long files that will take time, such as noise reducing an archive recording. It is inevitable that you will spend some time staring at the screen as the blue progress bar slides slowly from 0 to 100 per cent. Ironically, the time is much less than used to be taken with the mechanics of razorblade editing. However, it feels much longer because you, personally, are not doing anything while the process is taking place. You can improve your efficiency by having a number of other tasks that you can do while the processing is going on; telephone calls to make, facts to check or even letters to write, as the processing can carry on in the background while you use your word processor on the same PC.

Some routine operations, such as normalization, can often be done using 'batch' files. These allow you to leave the computer to get on with the task while you do something else (see page 70).

You will very often want to load many files at the same time. Please remember that the Windows file selector allows you to select many files, which will all be loaded when you click OK. One way of doing this is to use the 'rubber band' method of multiple selection (Figure 7.1). The mouse point is placed by a track and the left mouse button pressed and held down. With the left button still held down, moving the pointer will produce a rectangular 'rubber band' that will select any files included within it. You can include all the available files by starting the rubber band at the first file and then dragging downwards and to the right. The file selector window will scroll the list so that you can include everything.

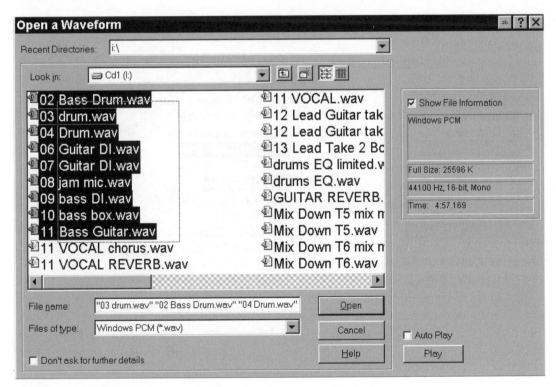

Figure 7.1 File selection using 'rubber band'

If you want to select a number of individual files, then click on each file you want (Figure 7.2). Normally selecting a new file will deselect the previous selection, but if you hold the control key down while you do this then this will not happen. The control key also introduces a toggle action, so that clicking on a selected file will deselect it. This is particularly useful if you want to make a minor modification to a number of files as, once loaded, switching between them is rapid. Once you have made all the changes, you can SAVE ALL and all your changes will be preserved.

Obviously, loading and saving a large number of files takes an appreciable time, but by triggering this with one action you can be doing something else that is useful while this is happening. If you are saving 10 files each taking 15 seconds to transfer, you have given yourself $2\frac{1}{2}$ minutes to make a telephone call or whatever.

Many programmes will include material from compact disc. This can be commercial music, but also sound effects or even material from your own personal archive. Here considerable time can be saved, as a properly set-up computer will be able to 'rip' the audio data from the CD much faster than real time. This means that 10 minutes of material can be transferred in less than 1 minute. You don't even have to listen to it! The quality will be better, as the audio is not converted back to analogue and then back to digital again by your sound card.

The easiest way is to use a background program that modifies the Windows desktop so that audio CDs appear in the desktop window with their tracks showing as .WAV files. These are usually

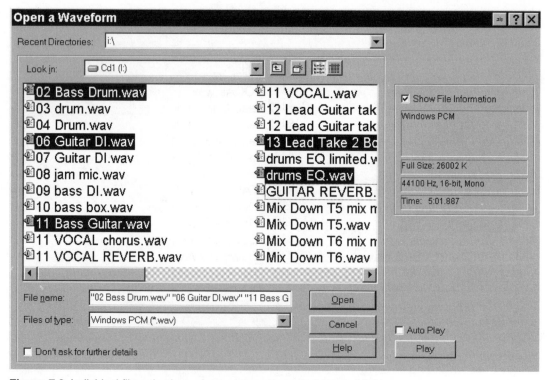

Figure 7.2 Individual file selection using mouse with control key held down

named 'track1.wav', 'track2.wav', etc. This allows you to copy the audio files as if they are normal data files. Later versions of Windows come with this feature built in, while earlier versions can have this added by either using a special file or a utility provided by the manufacturer of the CD-ROM drive. Windows desktops without this feature will open a CD drive containing audio, but the tracks will be shown as 'TRACK1.CDA' etc. and are not normally directly accessible as audio. Version 2 of *Cool Edit Pro* has CD Ripping software built-in, and can access the CD audio directly using the .CDA files as if they were the actual audio files.

7.3 Copy, cut and paste

Cool Edit Pro gives you the option of six separate clipboards – five of its own, plus the standard Windows clipboard. These are selected either through the 'Edit/Set current clipboard' menu option or by using control/1 to control/5 for *Cool Edit Pro*'s own clipboards and control/6 for the Windows one.

This means you can work with multiple pieces of audio 'in memory' at the same time – so you can, for example, copy different jingles or link music sections to each clipboard, and place them in your file at chosen locations. The current clipboard can also be set to be the Windows clipboard. This is available to other programs, and is a convenient way to copy audio from *Cool Edit Pro* to another program or *vice versa*.

These internal clipboards save audio in your temporary directory as wave files, and they can be retained even after *Cool Edit Pro* closes. The 'Delete clipboard files on exit' setting in Options/ Settings/System sets switches this on or off.

As well as using the normal paste function to insert material you can create a new file from the clipboard using Edit/Paste To New (Shift/Control/N). Save selection will usually be quicker, as you do not have to copy to the clipboard first.

Edit/Mix Paste (Shift/Control/V) will add material to the file on top of existing audio; the length is not changed except if the insert option is used. If the format of the waveform data on the clipboard differs from the format of the file it is being pasted into, *Cool Edit Pro* converts it before pasting.

There are four mix paste modes (Figure 7.3):

1 *Insert* – inserts the clipboard at the current location or selection, replacing any selected data. If no selection has been made, *Cool Edit Pro* inserts clipboard material at the cursor location, moving any existing data to the end of the inserted material.
2 *Overlap* – the clipboard wave does not replace the currently highlighted selection, but is mixed at the selected volume with the current waveform. If the clipboard waveform is longer than the current selection, the waveform will continue beyond the selection.
3 *Replace* – pastes the contents of the clipboard starting at the cursor location, and replaces the existing material thereafter for the duration of the clipboard data. For example, pasting 1 second of material will replace the 1 second after the cursor with the contents of the clipboard.
4 *Modulate* – modulates the clipboard data with the current waveform. I suspect that this option is here just because it can be done. It is yet another way of producing weird noises, and has the potential to do magical things. The classic use for this sort of facility is to make every day

sounds talk by modulating them with speech. However, the Dynamics Effects Transform can generate a modulation envelope. This can be used to modulate the level of a file as if it were being compressed using the audio from another file. This could provide 'voice-over' ducking for a music file.

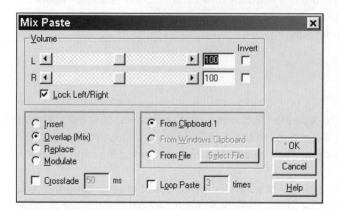

Figure 7.3 There are four mix paste modes

Loop paste allows you to multiply paste the same clip. This can be useful for music samples or for extending backgrounds and atmospheres for documentary work. You can choose to copy from the current *Cool Edit* clipboard, Windows or a file. The crossfade allows you to smooth the transition of the mix.

The invert tick boxes turn the waveform upside-down. Sometimes mixes work better if this is done. You can use this to compare nominally identical copies. If you make sure that the starts match, mixing the two, with one file inverted, *should* give total silence. Any noise or audio that is left corresponds to errors. It can be fascinating to do this when one of the files has been used with a lossy method; Minidisc for example. What you end up with is the audio that was 'thrown away'.

8

Structuring material

8.1 Standardize format

A simple way to split a large file is to mark it with regions corresponding to the separate files you want to obtain. This is done by using the mouse to select a section of audio and then pressing F8, which will create a 'basic' region (Figure 8.1). F8 combined with shift or control produces different types of regions used when burning CDs. There is also a 'Beat' marker that can be used when generating music loops using 'Edit\Auto Cue\Find beats and mark'.

You can trim each end of a region by dragging its marker to the right place. When you select View Cue List, you will then see your selections named 'Cue 1' upwards. Select the cues you want to turn into files. This is done in the same way as the Windows file selector. Hold down Control individually to

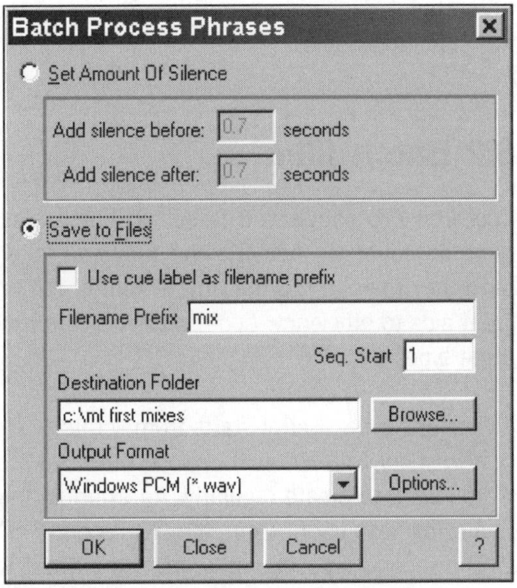

(a)

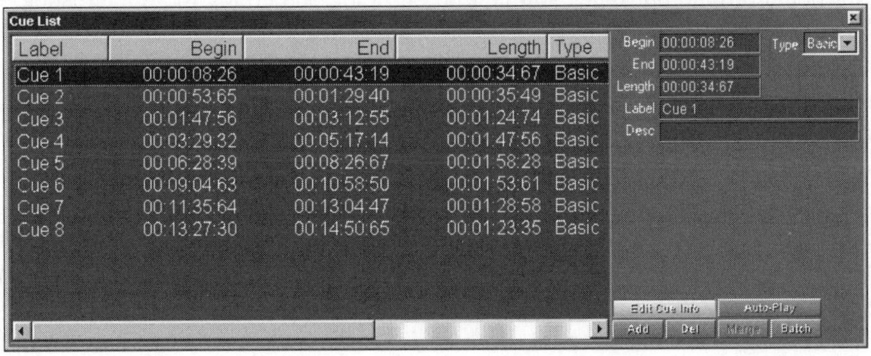

(b)

Figure 8.1 (a) Batch process phrases dialog used to split a long file into many small ones; (b) Cue list of regions

select a cue, or hold down Shift to select all the files from an initial selection to where you click. Press the batch button and you will see a new dialog.

There are two options. The first will insert a set amount of silence before each cue, and this can be useful to pad out the 0.7 seconds that are often missed at the start of MP3 files. In this case, we want the second option – 'Save to files'. Here you can set the type of file that will be output and to where on the hard disks on your system. The 'filename prefix' is what they will be named, with '01', '02' etc. added on to the end. In the illustration I have called the prefix 'Mix' and selected 'C:\MT First Mixes' as the output path. The resulting files will be 'C:\MT First Mixes\Mix01.wav', 'C:\MT First Mixes\Mix02.wav', 'C:\MT First Mixes\Mix03.wav' etc.

The files' cue names can be edited and these changed names can be output as file names by ticking the 'Use cue label as filename prefix' box. For example, I transfer several students' work from Minidisc in one go, and then divide up the different songs as regions, naming each cue with the student's name.

8.2 Batch files

Cool Edit Pro allows you to set up batch files to automate some activities. Using batch files is quite an advanced activity, and they can take a while to get right. This means that in the short term they may add to the time taken to edit an item. However, once they are sorted they can be great aids to efficiency. *Cool Edit Pro*'s Batch files contain sets of scripts that can be grouped into three types:

1 The first type, 'Script starts from scratch', creates a new file. The first command has to be File/ new (*Cool Edit Pro* will impose this if you forget). This is useful if you have a standard set-up you want to create regularly. An example might be to create a file at a particular sampling rate and bit resolution. Generate 20 seconds of line-up tone followed by 10 seconds of silence.
2 The second type, 'Script works on current wave', is where the same sequence of treatment is applied to the whole file and you can select a series of files to be actioned. This means that you can run a series of effects transforms that takes several hours to do in total, but instead of several hours of blue bar blues you can be off doing something more interesting – such as lunch!
3 The last type, 'Script works on highlighted selection', is one that is going to be applied to part of a file and consists of a series of treatments; say, equalization followed by normalization followed by 6 dB of hard limiting.

Scripts and batch processing can be found under the Options menu heading.

The scripts are, literally, text files containing instructions in a sort of programming language. But don't panic! You create scripts not by writing them, but by getting *Cool Edit Pro* to record what you do to a file. You can save this and recall it in the future.

To use an existing batch file or script, click on its name and then click 'Run Script' or 'Batch Run' (Figure 8.2). The 'Batch Run' button will only be enabled if the script is designed to operate on the whole file.

To record your own script it is simply a matter of entering a new name in the New Script title box, and this will enable the record button. The type of script you generate is controlled by your starting point in the editor: if there is no file, then the 'Start from Scratch' mode is assumed; if a section of the file is highlighted, then a script for action on a highlighted section is created; and a file with no selection creates the sort of script that can be applied to a series of different files.

Clicking 'record' starts the process. You now perform the actions you want to be able to repeat each

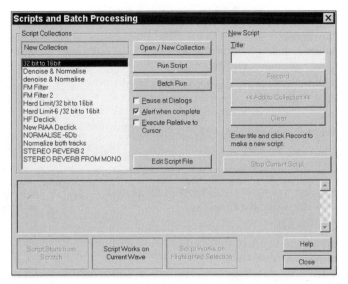

Figure 8.2 Scripts and batch processing dialog showing typical range of scripts

time the script is run. If you had ticked the 'Pause at dialogs' button, then when the script is run the dialogs will appear for you to enter settings (or press Enter to OK them). With the box not ticked, the values you enter as you record the script are used. It makes sense to use a short test file to create scripts, so you do not have long to wait for each action to be processed during the record process.

The 'Execute Relative to Cursor' is an advanced option, and is used when operating 'Works on Current Wave' script types. With this, you can have all script operations performed offset from your cursor position. If you record the script with the cursor set 10 seconds in from the beginning, then when the script is run it will operate 10 seconds after your current cursor position. Recording with the cursor at zero is a way of producing a script that works from the current cursor position, rather than using a highlighted section. This might be used for generating line-up tone sequences, for example.

As you would expect, the script recording is stopped by clicking 'Stop current script'. The '<Add to Collection<' button is now enabled so that you can save what you have done, or you can cast it to oblivion by clicking 'Clear'. This is the best option if you have made a mistake. For most people who are not techies, redoing the recording is preferable to clicking the 'Edit Script File' button, as this dumps you into Notepad and a raw text file that is the actual batch file of commands for the whole collection of scripts.

Where this can be useful is when you want a set of near identical scripts – for example, if you want to add a bit of silence to the front, normalize, and then add a variable amount of hard limiting. Figure 8.3 illustrates this with a script I have used for preparing speech recordings of audio books that are going to be transferred to MP3 files. I have four files that change the amount of hard limiting to give values of 6 dB, 8 dB, 12 dB and 16 dB of hard limiting. The scripts are identical except for the value on the line starting '2:', under 'Amplitude\hard limiting'.

71

Title: Silence Hard Limit 6dB
Description:
Mode: 2

Selected: none at 0 scaled 276973 SR 44100
cmd: Channel Both

Selected: none at 0 scaled 276973 SR 44100
cmd: Generate Silence
1:0.7

Selected: 0 to 307841 scaled 276973 SR 44100
cmd: Amplitude\Hard Limiting
1:0
2:6
3:7
4:100
5:1

End:

(a)

Title: Silence Hard Limit 8dB
Description:
Mode: 2

Selected: none at 0 scaled 276973 SR 44100
cmd: Channel Both

Selected: none at 0 scaled 276973 SR 44100
cmd: Generate Silence
1:0.7

Selected: 0 to 307841 scaled 276973 SR 44100
cmd: Amplitude\Hard Limiting
1:0
2:8
3:7
4:100
5:1

End:

(b)

Title: Silence Hard Limit 12dB
Description:
Mode: 2

Selected: none at 0 scaled 276973 SR 44100
cmd: Channel Both

Selected: none at 0 scaled 276973 SR 44100
cmd: Generate Silence
1:0.7

Selected: 0 to 307841 scaled 276973 SR 44100
cmd: Amplitude\Hard Limiting
1:0
2:12
3:7
4:100
5:1

End:

(c)

Title: Silence Hard Limit 16dB
Description:
Mode: 2

Selected: none at 0 scaled 276973 SR 44100
cmd: Channel Both

Selected: none at 0 scaled 276973 SR 44100
cmd: Generate Silence
1:0.7

Selected: 0 to 307841 scaled 276973 SR 44100
cmd: Amplitude\Hard Limiting
1:0
2:16
3:7
4:100
5:1

End:

(d)

Figure 8.3 Four files that change the amount of hard limiting to give values of (a) 6 dB; (b) 8 dB; (c) 12 dB; and (d) 16 dB

Rather than record four times, each with different values, I copied the section from 'Title:' to 'End:' and pasted it in three times to the text file. I then edited each section to have a different title for the script and change the hard limiting value.

Running the scripts in the future is a matter of going to the same dialog, selecting the sequence you want and clicking 'Run Script' for action on the current file, or 'Run Batch' if you want to process a number of files. The Batch Process dialog will appear. Click 'Add files' to get the file selector and add the files you want to the list (don't forget that the Windows file selector allows you to select more than one file at a time). If, having loaded a number of files, you see some that you realize you should not have included, then you can highlight them with the mouse and remove them by clicking the 'Remove' button (Figure 8.4).

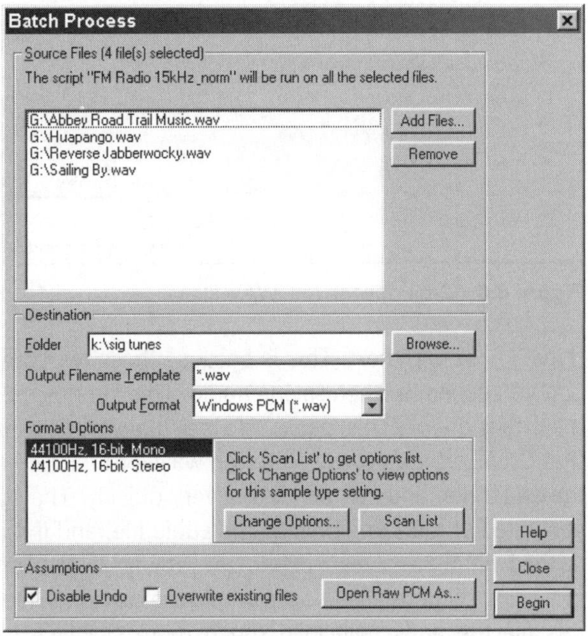

The directory box sets the path where the modified files will be put. This allows you to keep the originals, if

Figure 8.4 Adding and deleting files from the batch process

you have enough hard disk space. You can use different file formats, and they will all be saved in the format set in the 'Output format' box.

The Output Filename Template is an odd but powerful option if you can get your brain round it. The names of files in your batch can be modified before being saved to the Destination Directory. When running a batch, the processed file's extension will automatically change to that of the format chosen in Output Format (e.g. *.AIF for Apple AIFF). You can, however, force another extension, or alter the filename itself (the portion before the '.'), by using the filename template. There are two characters to use when altering the Output Filename Template: a question mark, '?', will signify that a character does not change, and a star, '*', will denote the entire original filename or entire original file extension (see Figure 8.5).

The scan list button will check all the files and tell what types you have selected. This can be useful in picking up a rogue file that has somehow appeared in the wrong format. The Change Options button allows you to

Filename	Template	Result
Zippy.aif	*.way	Zippy.way
Toads.pcm	Q* .voc	qtoads.voc
Funny.out	B???????.*	bunny.out
Boglong.wav	????.wav	bigl.wav
Bart.wav	* x.wav	bartx.wav

Figure 8.5 Here are some examples of how filenames (taken from the help file) will be saved given the original filename and the filename template.

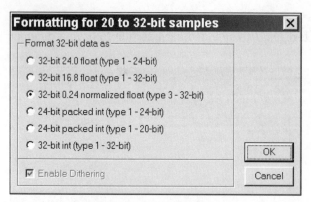

Figure 8.6 32-bit options for .WAV files

define how variants of the file format can be saved. For .WAV files, this sets how 32-bit data are saved (Figure 8.6).

The normal format is '32-bit 16.8 float (type 1 – 32bit)' This is the format *Cool Edit* is happiest with, and the only reason to change this is if the files are going to be used by a different piece of software that uses a different format. If the other software will not accept *Cool Edit*'s normal format, then try the second option '32-bit 0.24 normalized float (type 3 – 32-bit)' next. This is because these two formats can make use of a second file that *Cool Edit* can generate when saving audio. This file has the same name as the saved audio file, but has the extension .pk. This 'peaks' file is actually a specialized graphic file that contains the information that *Cool Edit* uses to generate the waveform picture on the screen. This file is much shorter than the audio file, and can be loaded very quickly. The software checks that the .pk file was actually generated at the same time as the audio file, and if it is valid then the 'picture' of the audio can be flashed onto the screen extremely rapidly. It also provides the software signposts as to where individual sections of audio are within the real audio file so that *Cool Edit* knows where to go. Without this file the audio will load much more slowly – try deleting a .pk file and then loading the corresponding file and you will see the difference. You will see *Cool Edit* scanning the file while it regenerates the file. Incidentally, because of this automatic regeneration you need not save the .pk files when you are archiving a project. The other formats have to be read in and the .pk graphics files created from scratch, which gives a slow loading time, whereas options 2 and 3 can use the .pk files to give a rapid loading time.

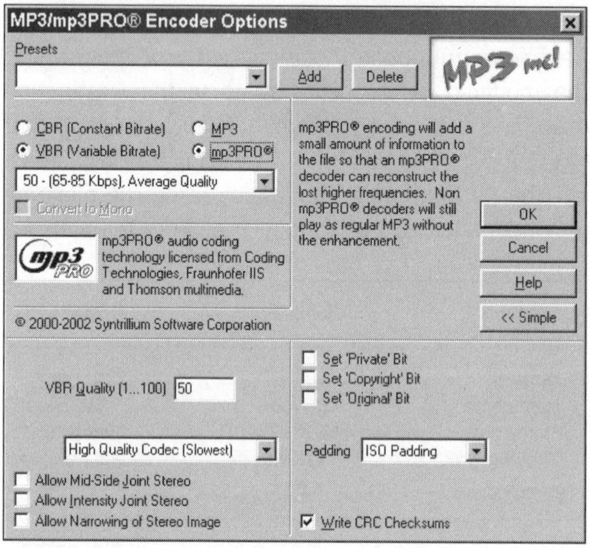

Other formats, such as MP3, have many more options (Figure 8.7).

The two tick boxes at the bottom of the scripts and batch processing dialog give important options. 'Disable UNDO' does just that and gives a great speed advantage, and is not even dangerous if you are saving the processed files to a different directory. The 'Overwrite existing files' box must be ticked if this is what you want to be able to do, otherwise the file will be skipped.

Figure 8.7 MP3 options dialog

9

Multitrack

9.1 Loading

Figure 9.1 illustrates the same four files loaded into three different multitrack editors; (a) *Cool Edit*, (b) *Cubase* and (c) *Pro Tools*. They have much in common, but differ in the way they operate and in appearance. In practice they are more similar than they are different, although this is not to say that they do not have different strengths and weaknesses, particularly in regard to the emphasis they give to MIDI (Musical Instrument Digital Interface). *Cubase* is a MIDI sequencer that can also handle audio, whereas *Cool Edit Pro* is an audio editor that can also handle MIDI. *Pro Tools* comes somewhere between the two.

At first, even a simple music balance such as in Figure 9.2 can look intimidating.

However, on initial loading a much simpler appearance is shown, as in Figure 9.3. Beside each of the four tracks shown are the individual controls for each track (see Figure 9.4). These are the default 'Vol' controls. Two other sets of controls can be selected by the 'EQ' and 'Bus' tabs at the top, and these are covered in Chapter 12 because they are concerned with controlling the real time effects available for each track. The 'Lock' and 'FX' buttons appear in each view, and are used to call up and hold FX. If the left-hand column is dragged wider then more controls will appear, giving a greater overlap between selections.

The essential ability of a multitrack recorder is that it can record on all of its tracks simultaneously, on only one, or on any number of them. So, in addition to the transport record button, each track has a red record enable button marked 'R'. If this is not selected then the track is in 'safe' mode and will not go into record when the main record button is selected. Additionally, already recorded tracks can be played back as a guide to the new recordings.

With multiple tracks, you are not restricted to stereo recording. *Cool Edit Pro* could record up to 128 tracks simultaneously, provided that the computer was equipped with adequate sound card inputs, fast enough processing and sufficient hard drive capability.

The button marked 'Out 1' is pressed to select which sound card output the track will be sent to. The '1' shows that it is being sent to the first stereo output in the list. The 'Rec 1' button controls where the record input comes from. Unlike some audio editors, *Cool Edit Pro* can handle both mono and stereo files. Others can only provide stereo by linking mono track together – if they import a stereo file, then that is split to two separate mono tracks. With *Cool Edit* you can use a mixture of mono and stereo tracks within the same session.

(a)

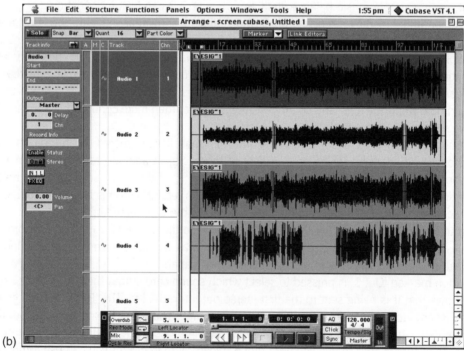

(b)

Figure 9.1 The same four files loaded into: (a) *Cool Edit Pro*; (b) *Cubase*;

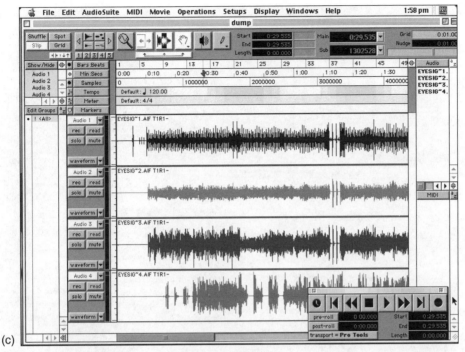

(c)

Figure 9.1 The same four files loaded into (c) *Pro Tools*

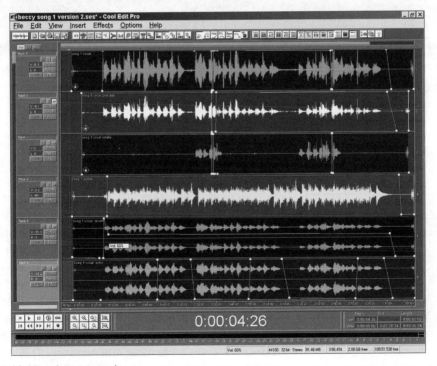

Figure 9.2 Multitrack start up view

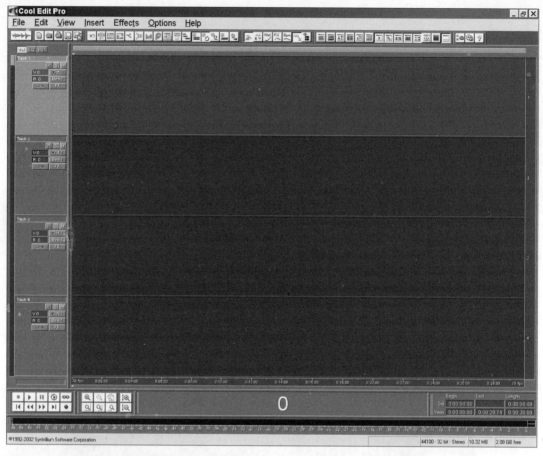

Figure 9.3 Multitrack opening screen

Figure 9.4 Track controls

For playback, each track, be it mono or stereo, is fed to a stereo output on the selected sound card (Figure 9.5(a)). A mono track is selected to only one output, by panning fully left (odd-numbered output) or fully right (even-numbered output). Although this may seem odd, this is fully in line with the convention used on multitrack music mixing consoles.

Each track can be individually panned between left and right. In the main window, this can be done in two ways. The first is to click on the Pan box (illustrated in Figure 9.4 showing 'Pan 0') and enter a value on the numeric keypad between 0 (centre) and 100 (fully right). Minus numbers will give you left pans, with fully left being –100. The alternative is to use the mouse to 'drag' the numbers. This is a method used by other editors, such as *Cubase* on Apple machines.

Place the mouse cursor over the Pan box, press and hold down the left mouse button, and drag it to the left to pan left or to the right to pan right.

The volume box (showing 'V0' in Figure 9.4) is altered in a similar way. Here the normal level is '0'. Volume is reduced when negative numbers are shown. Volume level is increased with positive numbers.

Right-clicking on the background of the box containing these controls gives you an actual slider to operate (Figure 9.5(b)). If you right-click on the track itself (not an audio waveform block), then clicking 'track properties' on the pop-up will give you the same dialog.

The record input can be fed from the left (odd) or right (even) card input or in stereo (Figure 9.6).

When starting a 'New session', you can also select either 16- or 32-bit recording. The sampling rate is fixed for the entire multitrack session. The 16-/32-bit option in the New Session dialog (Figure 9.7) selects what the mix down output will be. You can mix 16- and 32-bit wave files simultaneously. You can set up a default session that will preload wave blocks and settings that you have come to rely on.

Important though multiple ins and outs are, they are likely to be little used outside music mixing, where the multiplicity of outputs are useful mainly for feeds to an external mixer. For us, the power of the non-linear editor is the ease with which it can

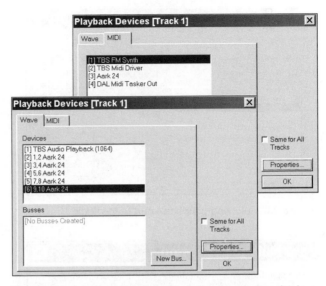

Figure 9.5 (a) *Cool Edit* selects devices by their order in a list. The 'Options/device order' menu item gives you a tabbed dialog for changing the device orders for playback, recording, MIDI in and MIDI out.

Figure 9.5 (b) Selecting track properties

merge and mix our mono and stereo audio into a continuous mono or stereo final mix. Our audio items are added and laid out, as required, on any number up to 128 of mono or stereo tracks. They

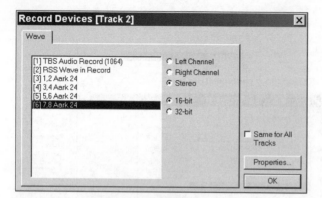

Figure 9.6 Selecting play back output for audio

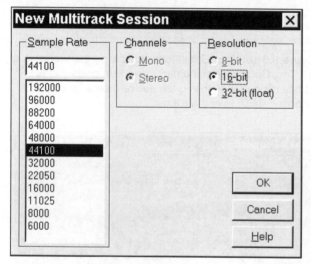

Figure 9.7 'New session' dialog

can be regarded as a large number of play-in machines in a studio. Without our intervention, all the material will be mixed together at full level with no panning.

It is time to look again at those tiny squares on the left of each track. We have already described the record button, the record and the playback selects. The two squares labelled with 'M' and 'S' are again mainly useful in music balancing. The 'M' mutes a track. This can be useful for switching off a track without losing all of its settings. The 'S' button is the solo facility. When selected, this mutes all the tracks except that one, and any others that also have their solos selected. Their pan settings are maintained, and so this is equivalent to 'solo in place' on a multi-track mixing console. You can solo more than one track at a time by clicking while holding down the control button.

9.2 Pan and volume

The two vital boxes are the 'P' and 'V' windows for panning and track volume. These can be operated in three ways: the first is to place the mouse cursor over the box and then drag it up or down to change the setting; the other two involve using a pop-up obtained by double-clicking on the box. The pop-ups each have a slider that can be dragged by the mouse, or moved by holding down the left-hand mouse button with the cursor placed on the slider on the side you want the 'knob' to move. Right-clicking on the slider will animate the knob, which will move to where you clicked.

Alternatively, the value can be edited in the box. Positive pan values can be up to 100 per cent (fully right), and negative values can go down to −100 per cent (fully left). These pop-ups set the overall level, and pan, of the track so that every item on it is affected by this.

Block

You can also set the volume and pan for individual blocks on a track. These may be separate wave files, or they may be wave files that have been split by the multitrack editor. These look like separate wave files but are, in fact, sections of a wave file. Setting is done by right-clicking on a block. Within the pop-up menu, 'Adjust wave volume' gives you a vertical slider that adjusts the volume of just that block (Figure 9.8a), while 'Adjust wave pan' gives a horizontal slider to adjust the pan (Figure 9.8b). The settings are shown in text at the start of the block with the letters 'V' and 'P', standing for volume and pan (Figure 9.9). These setting are in tandem with the track settings, and do not override them.

A good use for this is to make successive wave blocks consistent in level so that the master track level works for all the blocks. The block pan can be used to create a consistent pan setting for individual speakers in a documentary feature. If you split a wave block, the two sections carry the same volume and pan settings.

Envelopes

As well as these 'global' settings for the whole track or block, you can control volume and pan on a moment-by-moment basis. This is done using 'envelope' controls within the track. To be able to see the envelopes, you have to set the 'View/Show pan envelopes' and 'View/Show volume envelopes' menu options so that they are ticked.

Until changed, the pan envelope is a straight line down the centre of the track, and the volume envelope is a straight line along the top of the track. These represent pan centre and full volume (as modified by the track settings on the left). Clicking on a line will produce a small blob where you clicked. This is a 'handle',

(a)

(b)

Figure 9.8 Windows for (a) panning, and (b) track volume

which can be dragged by the mouse to change the setting. You must select the track by clicking it before this works.

To produce a simple fade-out, create a handle where you want the fade to start, and another one to the right of it. Drag this second handle to the bottom of the track (representing zero volume) and left or right to the point you wish the fade to finish. (You will not be allowed to drag it to before the start of the fade handle. However, this can be a useful 'end stop', if you just want to switch off the

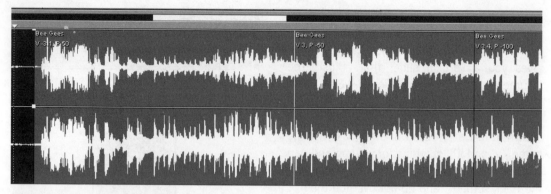

Figure 9.9 Three wave blocks with different volume and pan settings indicated by text at the top left of each book

track.) You will discover that the level immediately begins to increase after the fade, because the end of the envelope line is attached to the handle at the end of the track. This needs to be dragged to the bottom, so that the whole of the rest of the track is faded out.

Now you have your basic fade you can adjust it for timing by listening and, if necessary, moving the handles. A simple straight-line fade is often inadequate, but you can add as many intermediate handles as you like to make the fade curved, or to give it several sections (see section 9.6, Fades and edges).

You only need to use the pan envelopes where you want the sound from a track actually to be heard changing. You should beware of doing this to speech, as it is a distraction and reduces communication; the listener is likely to react on the lines of 'Oh they've started moving. Why? . . . er . . . what did they say?'

The pan envelopes are set in entirely the same way as the volume envelopes, although for much of the time a single pan setting for the overall track is all that is needed.

In both cases, a handle can be removed by dragging it off the top or the bottom of the track.

Figure 9.10a shows a short section of a mono track. This starts at full volume and ends with a fade. The pan setting starts at centre. Then there is a sudden pan full left followed by a slow pan from full left to full right. The next two sections are panned fully right and then fully left with the track ending panned centre.

When you want to move a handle, you position the mouse cursor over the handle until the cursor becomes a pointing finger. You can now drag the handle to where you want it, positioning it anywhere between the previous and next handles. You can see the value, which will pop-up beside the cursor (Figure 9.10b) It takes a little while to get used to this, and it is very easy to create new handles rather than moving an existing one. UNDO can correct this, or dragging the unwanted handle off the track will remove it.

This use of envelopes is a common way of overcoming the imposition of 'one-finger' operation by using a mouse. While you can only change one thing at a time, all the changes are remembered exactly. While simpler and less impressive than a screen filled with a pretty picture of a sound mixer, it can be much more effective.

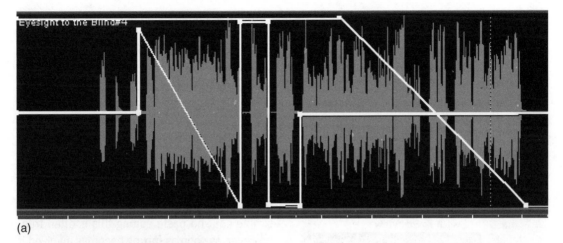

(a)

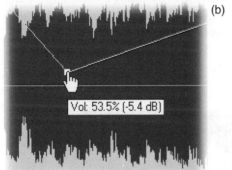

(b)

Vol: 53.5% [-5.4 dB]

Figure 9.10 (a) Moving envelope handle; (b) Producing a simple fade-out

To get rid of one-finger operation you need either an external mixer or an external control device that may look like a mixer but controls the software settings with the program. Many Digital Audio Workstations work this way, but at a price.

9.3 FX tracks

You now have complete control of the audio levels and timing of your material. Like many editors, *Cool Edit* allows real-time processing of audio, applying equalization and special effects such as reverberation. This requires a lot of processing and a correspondingly fast machine, but some sound cards have Digital Signal Processors that take the load off the computer's processor and have software to allow effects to be added in real time. Where this is not possible, the technique is to make additional tracks that contain the effects.

For example, to add reverberation to a soloist's voice, load the soloist's track (which is likely to be mono), convert it to a stereo track, and save it with a different name such as 'Soloist Reverb'. Now run a Reverb effects transform on a short section of your new track, and adjust the settings until you have the sound you want. Next set the 'dry' sound to nothing, and adjust the 'wet' sound so that it modulates the track at a sensible level. Now transform the whole track. This Reverb track can now be added in with the original when mixing (see section 12.5, Reverberation and echo).

Cool Edit Pro can also provide a 'halfway house' by 'locking' real-time effects. This is a way of pre-processing the real-time effects so that the processing power is released for other uses when mixing. Effects are more fully described in Chapter 12.

9.4 Multitrack for a simple mix

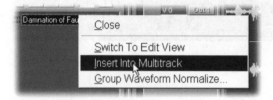

Figure 9.11 Using the 'Insert into Multitrack' option

It is very easy to see the non-linear multitrack mode as a complication to be avoided. For simple cut and paste editing this is often true; however, as soon as mixing becomes involved it is invaluable. It gives you the ease of multi-machine mixing with quarter-inch tape, combined with absolute repeatability and flexibility.

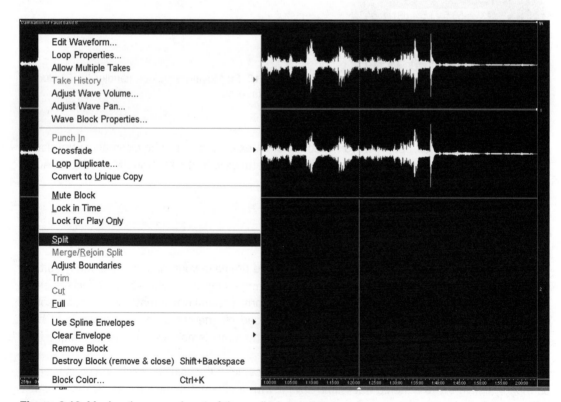

Figure 9.12 Moving the second part of the audio to track two

Because of the integrated nature of the editor, it can be convenient to 'pop in' to the multitrack editors when all that is required is to make a simple mix.

The following example takes a section of Berlioz's *Damnation of Faust*, which has one bar repeated in it. A simple cut edit does not work as a retake has been cut in that does not match the original. The answer is to cross-mix between the two takes during the one-bar overlap.

This is simply done by entering the multitrack mixer. If the wave file is already in the linear editor, it will already be in the editor's organizer list (Alt/9). All that needs to be done is to right-click it, and click the 'Insert into Multitrack' option (Figure 9.11). (To make the illustrations clearer, I have zoomed in vertically to show just two tracks.)

Next, the end of the first take of repeated section is selected with the mouse, and the track is split at that point. This is done by right-clicking on one side of the cursor and selecting the option to split the track (Figure 9.12).

Now that the audio is in two parts, the second half can be dragged, using the right mouse button, down to track two. In Figure 9.13 I have highlighted the first take of the repeated section.

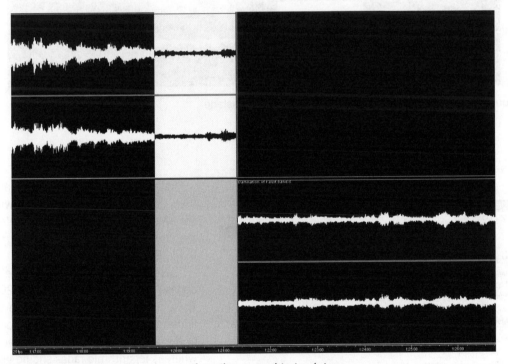

Figure 9.13 'Faust' file split and second section moved to track two

The next thing to be done is to slide the second section to the left until the overlapping section matches musically that of the track above (Figure 9.14).

At this stage, the rhythm matches but the level bumps and a crossfade needs to be implemented. This can be done manually with the volume envelope, but it is much easier to use the automatic crossfade facility.

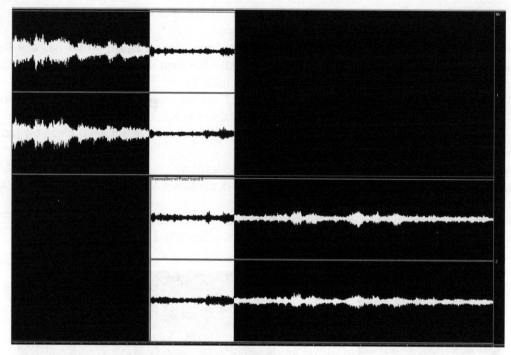

Figure 9.14 'Faust' slid to correct timing, ready for crossfade

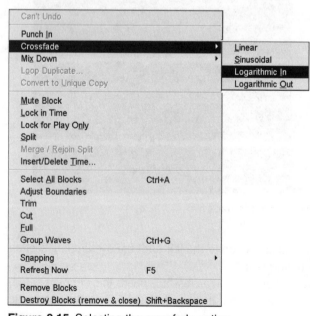

Figure 9.15 Selecting the crossfade option

First highlight the crossfade, and then both sections, by clicking with the shift key held down. Now select the crossfade from the menu (Figures 9.15, 9.16).

There are four types of crossfade to choose from (Figures 9.17–9.20). They all have their uses. However, linear crossfades tend to give a boost to the level in the middle of the fade, whereas logarithmic fades should give a better approximation to equal power thoughout the transition. Because of their asymmetrical shape, there is a choice of 'direction'. Sinusoidal is another equal power option. There is no substitute for trying the different options. The UNDO option is one

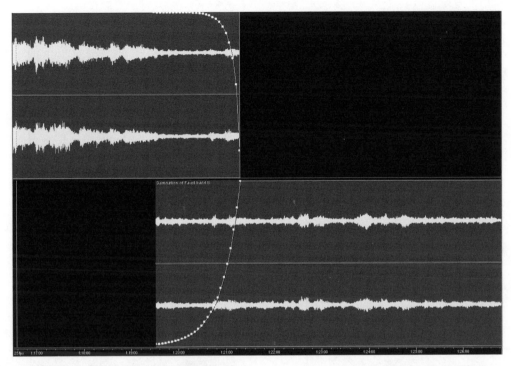

Figure 9.16 Crossfade has been implemented

menu click away (or ALT/Z). Playing the different transitions will soon convince you as to which is best for any edit.

To transfer this edit back to a file in the linear editor, you merely mix down your tracks and then save the file. If you are confident with your editing, you will select a short section with a little each side of the crossfade and mix down just that section. This can be copied and pasted in to your original in seconds.

Track bounce

There are times when you want to undertake a quick mix to make a composite track – for example, you may have two versions of a vocal track and want to mix the good bits together into a single track. You can 'track bounce' selected tracks by using the 'Mix Down to Empty Track N (Bounce)' menu option. N is the number of the next empty track. If you enable the 'Lock in time' option on the newly created wave block right-click menu, then it will stay in exact synchronization when you move it. This speeds up background mixing and helps clear up your work area, as you can now delete the originals from your session (although not, for safety, from your hard disk). The contents of all or selected enabled (unmuted) tracks are combined, with track and waveform properties (such as volume and pan) affecting the way the final mix sounds. Session elements such as looping, images and envelopes are all reflected in the mixed waveform.

PC Audio Editing

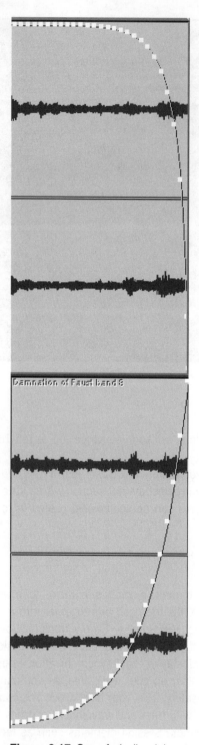

Figure 9.17 Crossfade (log in)

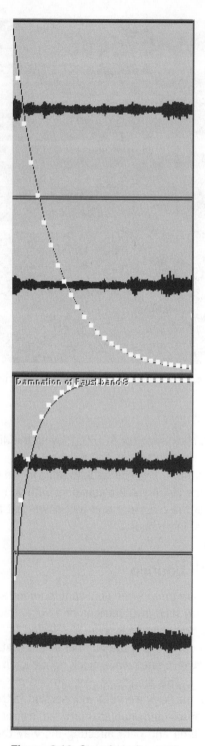

Figure 9.18 Crossfade (log out)

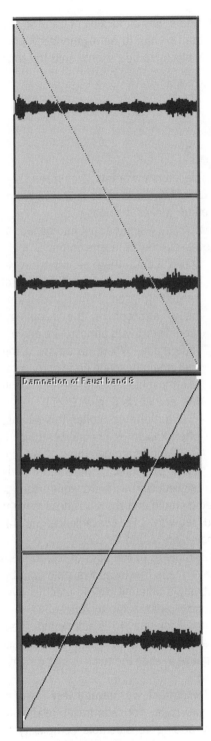

Figure 9.19 Crossfade (linear)

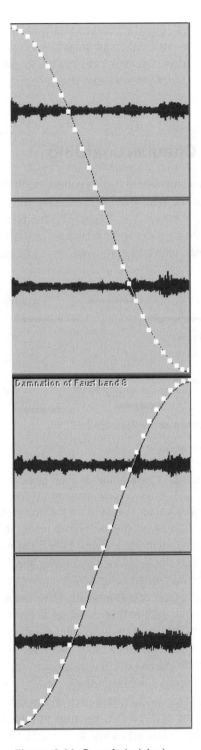

Figure 9.20 Crossfade (sine)

As track bounce also picks up your real-time FX it is also a way of 'locking' a track so that it is saved to hard disk, and so is quickly reloaded (rather than having to be regenerated) when you reload a session. It is not possible to select a locked FX block, so you should solo the track and choose the 'All Waves' sub menu option.

9.5 Chequerboarding

Once a programme item becomes much more than a simple interview or talk, then mixing becomes necessary. While mixing is possible with linear editors using Mix Paste, it usually involves a lot of UNDO cycles to get it right. The multitrack non-linear editor makes it so easy.

In the days of reel-to-reel tape, programmes of any complexity would be split into banded 'A' and 'B' reels, with odd-numbered inserts on the 'A' reel and even-numbered inserts on the 'B' reel. Each insert could then be started over the fading atmosphere at the end of the preceding insert. Where both items had heavy atmosphere, the incoming band would start with atmosphere and a cross-fade made. Where necessary, a third 'C' reel would be made up to carry continuity of atmosphere over inserts with tightly cut INs and OUTs (Figure 9.21).

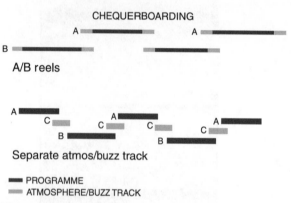

Figure 9.21 'Chequerboarding' on tape.

In a multitrack editor, this technique is often known as chequerboarding. Figure 9.22 shows a programme of music linked by a presenter. Some of the music is segued, and so the chequerboard principle is invaluable. (*Segue* is an Italian music term adopted by radio and show business to mean one item following another without a break. It is pronounced 'seg-way'.) For convenience, the links are given their own track (track 1) and the music tracks their own 'A' and 'B' reels.

A degree of extravagance in allocating tracks is not a problem. In the days of reel-to-reel tape, studios were often limited by the number of physical play-in tape machines available (and by how good the tape operator was!). Here, each track is effectively a different play-in machine. We can have up to 128 of them – rather better than a four-machine radio studio (Figure 9.22).

Even better, we are no longer reliant on the tape operator. Each 'play-in' is automated; fixed until we change it. No longer does one muffed play-in require a whole take to be redone. If an item comes in slightly late or early, it only takes a moment with the mouse to slide the track until it is right (drag using the right mouse button).

While totally dominant in music recording, reel-to-reel multitrack was never a very useful device for drama and features because of this one limitation; you could not slide tracks relative to one another. There are ways to emulate this, but only by adding considerable complexity to the process.

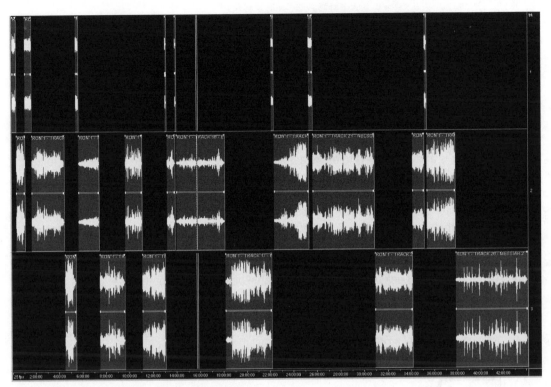

Figure 9.22 'Chequerboarding' on *Cool Edit Pro*

9.6 Fades and edges

However, a little more than simple chequerboarding is required. Together, a studio manager on the mixing console and a good tape operator would make material merge seamlessly by skilled and instinctive use of faders, both on the console and on the play-in machines. Often inserts would be started not with an actual fade-in but with edge-ins, where the item was started with a sharp fade-up to 'half fader', and then a fade-in over a second. Sound balancers have also learned that most items sound better if they are started 2 dB to 4 dB low, and edged up over a second or so.

On a PC, all these instinctive skills have now to be defined. This is why many people, with the necessary money, will pay for an external mixing device to control the computer's mixer. This preserves the power to adjust levels minutely within an item merely by having your fingers on the faders. This 'pinky-power' can control up to 10 sources simultaneously, whereas a computer mouse can only control one. However, because of the automation provided by a PC, each change can be made individually and cumulatively.

It makes sense to preserve the old split of duties between the sound mixer and the tape operator. When preparing your material, pay attention to how each file begins and ends; put a rapid fade-out on the atmosphere at the very end of the file. Put an edge-in on the front. You can set this up as a 'Favorite'. Starting values of −6 dB and −3 dB fading up to zero are useful here.

Edge-ins should be inaudible to the listener. They are there to smooth the seams. Actual fade-ins and fade-outs – ones that are going to be perceived as such by the listener – are best done using the envelope controls of the multitrack mixer.

Fades usually have the role of establishing a passage of time. The old convention, from the 'golden age of the wireless', was a slow fade-out followed by a slow fade-in. These days, the fades are much faster and the fade-in is more of a steep edge-in.

Fades also have the function of allowing voice-overs to be heard over music. Very rarely should the listener be aware that this is being done. What will sound particularly horrible will be a mix where the music is audibly dipped to make a hole for the voice to enter, with another hole following the end of speech before the music is faded back up.

Most important is that the voice-over should fit the music. The music should return at the start of a musical sentence. The dip-down is less critical in the context of pop music, but with more formal features the voice should pick up from a cadence, or the end of the musical sentence.

When mixing voice-overs 'live', I used to say to presenters that I would 'go on the whites of your tonsils'. By this I meant that the music would dip on the very first syllable of their link. Equally, it sounds better if the fade back up starts about a syllable before the end of the link. This is less critical if the presenter has hit the beginning of a musical phrase as you are fading up in the micro-pause between notes.

Notional Structure of a Fade-out

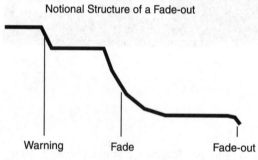

Warning   Fade   Fade-out

Figure 9.23 A fade-out consists of three parts

Music fade-outs are often done badly. A vital thing to realize is that any fade-out consists of three separate parts (Figure 9.23):

1 Warning listeners they are about to lose the music
2 Fading down
3 Fade-out.

This applies as much to a clean fade as to one where the presenter comes in over the end. The classic example of the latter is the opening signature tune, a feature of so many audio items, from broadcast programmes to cassette promotions. The music starts full level, and then towards the end of a musical phrase the level is dipped slightly to warn the listener, followed almost immediately by a fade-down so that the presenter can start his or her link at the end of the phrase with the music running under. The music is then lost under the presenter, and brought out on a cadence, or end of phrase.

Fading classical music badly can cause real offence, but pop music deserves equal care. Again the three-part fade applies; a slight dip at the start of the last phrase and then a fade towards the closing cadence, with the music taken out at the end of the cadence. A common mistake – especially when the fade is made while looking at the score – is to leave the start of the fade too late so that the listener is not prepared for the music to end.

Where the fade has to be quick, say because the illustration is about the words of a song, then, rather than lose the last few words on a fade to cadence, a 4–6 dB dip at the beginning of the last line can provide the psychological warning and make the quick exit far less offensive.

9.7 Prefaded and backtimed music

A common technique to bringing a programme out on time – especially a live one – is to have 'prefaded' closing music. A piece of music is dubbed off, say of 1 minute's duration, and started with the fader shut exactly 1 minute before the end of the programme. As the presenter makes the closing remarks, the music is faded up behind the words to bring the programme to a rousing and punctual close.

A technique used by some producers is to end the last item with music, which is then run under the presenter's close and then faded up so that it ends. This effectively offsets the prefade back to the last item. However, it sounds so much better if that last piece of music actually begins at a musically satisfying place, rather than just emerging arbitrarily. A cheat that works most of the time – although I remain offended that it does – is to dub off the item's closing music to the end, and run that separately as a backtimed prefade closing music. The two recordings will not be in synch, yet because they are the same music they can be crossfaded under the presenter's closing link without being audible to the listener.

These techniques may seem unnecessary for a recorded programme, as it can be edited to time. However, time to edit can be a scarce resource, and it may be much more cost-effective to use prefaded closing music than to spend another hour looking for cuts. Of course, you do not have to do any actual backtiming. All you have to do is slide your closing music until its end is at the duration that you want. You then use the amplitude envelope control to fade it in at the right point.

9.8 Transitions

Transitions are the stuff of dramatic effect – not only in drama itself, but also in documentary. Crossfading from one acoustic environment to another or from one treatment to another can be extraordinarily effective.

The first type of transition, going from one location to another, is simply managed. Two tracks are used, and the incoming track slid along to overlap the end of the exiting track. You can use *Cool Edit Pro*'s automatic crossfade generation, as described earlier in this chapter, or you can manually create the envelopes if this does not produce the effect you want. Always remember that you can modify the crossfade that *Cool Edit Pro* did for you, by moving the envelope 'handles' with the mouse. With a transition, you are rarely going to use a perfectly engineered equal loudness crossfade.

The second type of transition is from one type of treatment to another on the same material. Examples include an orchestra playing in another room. The music is treated to remove all the high frequencies. The characters move into that room, and you want the high frequencies to return as they make that move, or they open the door. Sometimes you want to go from music or speech apparently heard acoustically – a juke box playing in a coffee bar – to music heard directly off the recording. Here you need to take the good quality original and treat it, literally by playing it on a loudspeaker and recording it with a microphone (or, in a studio, playing the music on the foldback loudspeaker). A small room effect may be used instead.

With the example of the orchestra, your treated track is identical in length with the original and is inherently in synch with it. Within the multitrack editor, place the cursor roughly where you want the treated track and, without moving the cursor, insert it. They will both line up on the cursor. Now they need to be grouped together so that you can move them without losing synchronization. Click the first track so that it is selected and any other selection lost. Now, holding down the control button, click the other. They should now both be selected. Right-click on one of the waves, and click group waves on the pop-up. You can do this from the 'Edit/Group waves' menu item as well. Now when you move them about they will move together, until you match the dialogue in the best way.

You now take the handles at each end of the volume envelope on the good quality track and move them to the bottom, so that there is no output from the track. Listening to the mixed output, adjust the track gain for the treated track so that it fits behind the dialogue. If it is supposed to be coming from behind a door, then you may wish to pan the track to one side or the other.

At the point in the dialogue where you wish to make the transition, create a crossfade by adjusting the envelopes on both waveforms. You can adjust the gain of the incoming untreated track with the track volume control, or by using the volume envelope.

In the second example, the music recorded off a loudspeaker, the two files will be of different lengths. However, it is very easy to slide one file with respect to the other until you can hear that they are in synch. Now you can group the two files and treat the transition in the same way as before.

9.9 Multitrack music

Music recorded on a multitrack recorder can be transferred to your computer very easily, if you have a multitrack sound card with enough inputs. This can even be done digitally with the right gear. A common interface is a variant on the optical TOSLINK connector. The same plug is used, but an eight-channel digital signal is sent instead. An increasingly common type of multitrack recorder is the video cassette-based machine, which records eight digital audio tracks instead of a picture. These can be linked together to become 16-, 24-, 32-track etc. recorders.

In cases of desperation, short pieces of music – say up to 5 minutes – can be transferred using only stereo equipment. The multitrack tape must be prepared by having a synchronizing 'slate' recorded simultaneously on all of the tracks just before the music. A convenient way of doing this in a studio is to use the slate talkback – the talkback that goes to the multitrack and not to the artists. A short ident, and a sharp sound like a coin tapping the talkback mic, should be adequate. At the end of the music, a similar slate is recorded. The tracks can now be copied two at a time, either directly to the computer or to a stereo recorder, preferably digital. When the tracks have been transferred to your computer, they should be split into their corresponding mono files. This will have been done automatically if you have used the multitrack facility to record them as two mono files rather than a single stereo file. Each wave file will begin with identical recordings of the slate. Each file is edited so that it starts with the most recognizable part of the 'sharp' sound. Sample accuracy editing is advisable (and practical) here.

Once loaded into the multitrack editor and lined up at 0:00:00:00, the files should now be back in synch. How much error has been introduced can be ascertained by examining the end slates.

Usually, in my experience, the error, even with an analogue multitrack, is better than two EBU frames (2/25th second), which for many types of music (and performance) is good enough. Better accuracy can be obtain by using *Cool Edit Pro*'s effects transforms to varispeed tracks digitally so that the error is corrected. The mix down can now be undertaken within *Cool Edit Pro*.

MIDI and video

Both MIDI and AVI video files can be loaded in to the multitrack file and can be used as references for audio tracks (Figure 9.24). The picture is represented by a graphic of 35 mm film on the multitrack display itself. Its track has no operational controls at the left. The actual picture can be seen in a separate window by clicking 'View/Show Video Window'. The audio has the full range of operational controls.

A MIDI file's notes are represented graphically, but this has only the mute and solo buttons available at the left. There is an additional 'Map' button that allows you to adjust the MIDI device and channel assignments. The MIDI part also has a volume envelope for adjusting it within the mix.

Figure 9.24 AVI video and a MIDI file open in *Cool Edit Pro*

Loops

Wav files can be configured as loops. To do this, right-click the block and select 'Loop properties'. Once so configured, they can be dragged out to repeat for as long as required. The prime function for this is to use samples of drumbeats etc. to make up a rhythmic background for music balances. For speech and drama work this can be very useful for looping small sections background effects or actuality to extend them. However, beware that short loops soon set up their own rhythm. Longer loops can also become recognizable if run for a long time. For example, a horn sounding in a traffic noise loop will become startlingly obvious the third or fourth time around!

10

Production

10.1 Introduction

This is where the disparate strands of material you have been gathering get pulled together and made into your final programme, or programme item. The script is written, or a running order prepared. Each item is put into the correct order, checked with the existing material and recorded. Traditionally this was done as a studio session. While there are still plenty of occasions when this is still the case, the digital audio editor can make this less necessary.

With many speech-based programmes, the main purpose of the studio is to provide a quiet (and relatively dead) environment for your presenter to record voice links between items (*aka* inserts) and voice-overs on top of actuality. Given that your presenter is in this sterile environment, it makes sense that he or she should be able to hear the insert material in context so that it is possible to react to it.

However, small, cheap and light, portable recorders make the studio increasingly unnecessary. Its remaining function of providing a sound console to mix the material together seamlessly is subverted by the digital audio editor.

10.2 Types of programme

There are many types of recorded programme, from a simple talk to a full-scale drama or music recording, and a modern PC has within it the resources to cope with all of these. The only programmes not within its scope are, arguably, those that need two-way communications. The interview on the telephone or circuit is still best done in a studio, with its specialist communications and talkbacks. While a PC can handle a modern ISDN line, it is the sophisticated communication between the producer and the interviewer, as well as the two-way discussion on the circuit, that is important. While it is not impossible for a computer to do this, the requirement is sufficiently specialized as not to be something that you can buy easily or cheaply.

Incidentally, when time is available, there is a technique for recording telephones in good quality even when international or mobile calls are involved. Put simply, each end of the conversation is recorded locally, in good quality, on a portable recorder (or, of course, in a studio). The two tapes or discs are then combined in the digital editor after they have been posted, or freighted, to bring the two halves together. With digital recorders little slippage of synch will take place, but even when it does a little sliding of tracks can easily be done. A track can always be split while it is silent, because 'its' person is listening to the other end. The best results are achieved if the room atmospheres at each end are kept running continuously. The main difficulty with this technique is that people tend to speak differently on the phone. On a poor line they shout, and this sounds very odd when both voices sound as if they are in the same room.

10.3 Talks

The most straightforward, if not easiest, production is the straight talk. Once common on radio, this now struggles to keep in view. However, it also manifests itself as commercial promotions – a sales message on cassette, or the chairman's morale-boosting address to the company's employees.

This merges into the illustrated talk, which can soon become a magazine programme.

From the production point of view, talks tend to be serial in nature. They start at the beginning and finish at the end. Talks will be all the more effective if they are not overburdened with detail. In the 1950s, the BBC Radio Talks Department thought that the ideal talk would handle one concept in 15 minutes, provided that the structure was of the form:

In this talk I shall tell you X.
I am telling you X.
I have just told you X.

This may be effective in communicating X, but it is difficult for it to be entertaining. Somewhere between the chairman's address and the above will be what you want. Do you want the audience to retain the facts, or to get a good impression of your chairman?

There are two likely locations for this talk; a special recording session or a public meeting. In the first case you are in control; in the second you are not. Do not be tempted to use your best music mics, as they are almost certain to be a popping disaster. A good moving-coil mic that is known to be insensitive to making popping and blasting noises is best. It is also unlikely to have an extended bass response that will pick up every low frequency distraction, from distant slamming doors to air-conditioning noises.

Will the speaker stand or sit? When recording a public meeting, ignore what the speaker's office tells you and use two microphones. Set one up for the speaker sitting down, and rig the other for the speaker standing up. If at all possible, use floor stands for the mics so as not to pick up table thumps. Place the mics about 60 cm (2 feet) from the speaker's head; any closer and the balance will become very sensitive to head and body movements. The further away the mics are, the less likely speakers are to move them. (There is a perception problem here; if a mic is very close and is pointing at their lips, speakers have an urge to move it so that it is 'pointing at them'. This leaves the mic pointing at their eyes. This in itself is not a problem, but their desire to move the mic is. 'Miking' the eyes can, therefore, make sense.)

If a mixer is not available, then splitters can be purchased that will allow the two mono mics to be fed separately, one to the left-hand channel and the other to the right-hand channel of a portable recorder.

If you want audience reactions – applause, etc. – then you need an additional pair of mics left and right of the audience. These should be fed to the mixer, if you are using one, or to a second recorder (two Minidisc recorders are more generally useful and cheaper than a decent mixer). If both of the recorders are digital, there will be no problem with synchronizing the recordings once they are transferred to the computer.

Place the microphone for secondary speakers slightly to one side but pointing at them. They will tend to talk towards 'the boss', no matter how many times they have been asked to talk to the audience! You will obviously need a mixer for this sort of event.

For the specially recorded talk, you are more in control. Someone sitting at a desk, with a table stand holding the microphone, will produce adequate results from an audio point of view but, very likely, a very formal delivery. This may be what is required.

Another option is to record the talk as if you were recording an interview, but without any questions being asked. The speaker has eye contact and is more likely to talk to you, representing the idealized single listener. The vital thing to remember with all recorded (and broadcast material) is that you are communicating with individuals. There may be thousands of them, but they are not a mass audience; they are a large collection of individuals listening by themselves.

10.4 Illustrated talks

Best results are obtained if the illustrations can be played in as the talk is recorded. This allows speakers to react to the material more naturally than if they record 'cold'. Whether or not the inserts, as played in, are recorded at the time is down to production convenience. A minimalist set-up might be two DAT or Minidisc recorders, one being the play-in machine, perhaps feeding a couple of small powered speakers as used for PC sound.

10.5 Magazine programmes

Structurally, the magazine programme is a series of premixed items linked by, usually, one or two presenters. In many ways it is rather like an illustrated talk but on a larger scale. The individual items are often 'presented' by reporters and contributors other than the main programme presenters.

10.6 Magazine items

Magazine items often take the form of illustrated talks, with a presenter/insert/presenter/insert format, but they can also be simple features.

10.7 Simple features

A simple feature differs from an illustrated talk by being continuous, with an attempt to make the item seamless. The presenter will talk over actuality, and interviews will be chopped up and presented as extracts.

10.8 Multi-layer features

The multi-layer feature is the most complex form of programme, where various threads are interwoven to make a seamless whole. Before computer audio editing, these were time consuming, labour intensive and complex to make. The multitrack non-linear editor doesn't remove the complexity, but does make it much more easily handled. Expect at times to be mixing a dozen tracks simultaneously.

10.9 Drama

Drama brings with it the widest range of techniques. It can be formal and theatrical when reproducing a stage play, and it can also be shot like a film, using portable equipment on location.

Drama is often studio-bound for the simple reason that, with no cameras, the actors are traditionally not expected to learn their lines but to read them off scripts. There is often a technician in the studio to do 'spot' effects, such as pouring tea and knocking on doors. This is the radio equivalent of the Foley artist used in cinema films.

Each speech in a drama script should have a left-hand margin with the character name at the start of each speech, which should also be numbered, starting from 1 on each page. This allows speedy and simple communication by referring to 'Cue N on Page Y'.

Drama often (but not inevitably) leads to a lot of post-production. Background effects are best done at the time, as it is especially valuable to the actors if they can hear what they are talking against. This is the advantage of location drama, as the sounds are real. If the actors are on a street, they raise their voices quite naturally as a bus passes. This is also a problem with location drama, as you will probably have no control over that bus passing. It can also result in tiring the listener, unless there are contrasting scenes away from the hurly-burly.

Modern technology allows drama that is set in modern times to be made quite effectively with portable equipment and a PC for post-production. The biggest problem can be making it clear to passers-by what is happening. People are now quite blasé about camera crews, but can be quite confused when there is no camera. I have even heard the suggestion that a balsa-wood mock-up camera would be an asset! The idea is to prevent your best take being ruined by the microphone picking up passers-by speculating amongst themselves as to what all these strange people are doing!

10.10 Music

This encompasses the whole range, from the solo singer to the multitracked band to the symphony orchestra. There is also the range from electronically synthesized and sampled music, usually using MIDI, to acoustically produced sound from traditional instruments. Music technology is a whole separate industry and is beyond the scope of this book, except to say that, surprisingly, little has to be added to a PC already equipped for audio in order for it to take on the full gamut of music.

Cool Edit Pro has the capability of importing MIDI files as tracks to the multitrack mixer. Some basic simple editing is also possible. This means that your music lovingly prepared on a sequencing package like *Cubase* can be run into your synthesizer as part of the *Cool Edit* session. An alternative solution is to synchronize *Cool Edit* to your sequencer using MIDI Time Code.

11

Post-production

11.1 Timing

This covers the period between the compilation of an item – often in a studio – and the final mastering of the finished item. In some cases it is difficult to distinguish this phase separately, but it is rare for it not to exist at all.

With a 'live' transmission, it is unusual for adjustments not to have to be made after a rehearsal. Only in live news and current affairs programmes do adjustments have to be made on-air. An overall studio producer/director will ask colleagues to cut arbitrary amounts of time out of items. Hellish though this can be, it usually works.

Typically, when all the sections of an item have been put together, whether in a studio or with your audio editor, the item will be too long and poorly paced, with further editing of sections required either for style and pace or for 'fluffs' and overlaps. It is always a waste of time to 'de-um' speech recordings until this stage, as otherwise you waste time on edits that are thrown away.

Don't regard it as a 'mistake' when you find your post-production material is too long. It's difficult to judge the worth of individual items until they are in their final context. Something that seemed relevant within an individual item becomes a drag within a programme. Ten per cent over-recording is about right. Now is also the time to be rigorous with yourself.

A good traditional policy when looking for something to cut is 'Cut your darlings'. That piece that was particularly difficult to get, that fact that is so fascinating to you, will, as likely as not, turn out to be supremely indulgent and weaken an otherwise tightly crafted item.

The post-production phase is also where you apply your audio design, which is covered in Chapter 12.

11.2 Level matching

Make sure that levels are matched. The human ear is extremely sensitive to quick changes in audio levels, but it is much less sensitive to slow changes of level. This is, at once, a danger and a boon.

The danger is that, unless you are constantly keeping an eye on levels, they can slowly sink. There is a tendency for the ear to want successive sounds to be about 2 dB quieter for the level to sound matched. This is why the edge-in is so important in smoothing transitions. You can have the level the ear expects and, by slowly increasing the level, restore it to normal.

An edge-in involves a small change in level, but sometimes original material has far too much dynamic range. An obvious example might be classical symphonic music, where the range from ppp to fff can easily exceed 50 dB. The traditional BBC guideline for dynamic range was a mere 26 dB, and this is still valid for an item that someone is going to sit down and listen to. However, it is extraordinarily rare for anyone to do that. Radio broadcasting and audio presentations have the advantage that they can be absorbed by people who are busy, or on the move. In the modern noisy environment, a dynamic range of 12 dB is about as wide as you can go without audibility suffering. Real life has a dynamic range of 120 dB, a power ratio of 1 000 000 000 000 : 1. How can we represent it with 12 dB, or only 16 : 1? The answer is, by manipulating the ear's strength and weakness.

A fully modulated speech programme item will consistently peak to full level. Dynamic range is implied by cheating levels at transitions – a slow change of level is not perceived; a sudden change is. You can make an entry sound loud, be it a fortissimo passage or a shout, by dipping the level before the transition.

Say we have a presented item that is back-announcing a previous item on Mozart, followed by music played by The Who or The Rolling Stones. This is meant as a surprise. It is meant to sound loud. The presenter then voices over the music and must now sound louder than the music. Here, the answer is to creep down the level of the back announcement and start the music at full level. This is crept down and then dipped conventionally for the presenter's voice-over, which is now at full level. This happy jig is kept up throughout the programme, giving a sense of dynamic range yet keeping the programme audible and intelligible, even when heard in a car on a motorway.

While broadcasters and makers of CDs usually have no control over the environment where the listener will hear their recording, there are exceptions. Examples include *Son et Lumi'ere* and tourist guides on cassette or CD heard on headphones in museums, monuments and historic buildings. At places like these, a wide dynamic range can be part of the entertainment.

11.3 Panning

There is a school of thought that says that all speech should be in mono in the centre, and only music and effects should be heard in stereo. This may seem a little severe until you realize that all modern cinema films are balanced this way, as is much of the stereo sound on TV programmes.

In practice, speech inserts into programmes are in mono because they are easier to edit, if the image is not constantly moving around. A common convention is to put the presenter into the centre and then put contributors in panned 6 dB left or right. Panning more than this becomes gimmicky and potentially reduces compatibility if heard in mono; a contributor panned too far in either direction will sound quieter than the central presenter. Another convention is to aim for a symmetrical balance, with interviews panned equally left and right, or discussions panned symmetrically across the sound stage. (Studio interviews, using the presenter, should be recorded with the contributor panned hard left and the presenter panned hard right, using separate microphones. This gives total flexibility of positioning in production.)

11.4 Crossfading

Crossfades have several functions:

- Equal power crossfades hide the join between separate, but similar, sections of audio
- A dip in level combined with a change in sound can be used to indicate a long passage of time/ substantial change of location; this is effectively a modern contracted form of the old fade-out, pause, fade-in convention
- A simple change of sound without level dip can be used to indicate a short passage of time and a small change of location (say, to get from one room to another).

11.5 Stereo/binaural

Conventionally, audio is heard on loudspeakers. Normally there are two of them, but for surround sound systems like Dolby Stereo there will be more. These surround systems either use special coders/decoders to play phase tricks with two-channel sound, or they use multiple channels. DVD has standardized on a 5.1 channel surround system. The '.1' is an engineer's jokey way of saying that the sixth channel is low bandwidth, suitable only for conveying low frequency audio (earthquakes, spaceships, etc.).

Binaural sound has been around since the beginning of stereo, when Clément Adèr relayed the Paris Opera to an exhibition in 1881. Visitors to the Paris Exhibition listened using two earpieces fed separately from different microphones at the theatre.

Techniques for recording binaural sound usually involve some form of dummy head. This can range from an actual model of a human head with microphones in its ears to something more stylized but just as effective; a Perspex disc approximately the size of a head, with small omnidirectional microphones mounted either side about 22 cm (9 inches) apart. (The relative transparency of the Perspex disc makes it acceptable for slinging above orchestras at public concerts; a disembodied head might disturb the audience!) There are also microphones designed to fit in a person's ears, and thus a real head can be used. This does have the disadvantage that it is very difficult for the person wearing the microphones not to turn his or her head towards different sounds, which produces a very disturbing sound image for the listener.

Binaural seeks to emulate how we actually hear. The resulting sound is surprisingly good heard on loudspeakers, and can be dramatically effective heard on headphones (open headphones work better than enclosed ones). When it works for you, a totally three-dimensional sound is heard. However, it is said that only 60 per cent of the population can get the best out of the system. A large proportion of the rest get a 3D effect, but not one emulating real life. A common problem is everything seeming to be coming from behind the listener.

It works using phase and frequency response variations caused by the obstruction represented by the dummy head. This gives sounds outside the listener's head. For example, a mono voice panned fully left, heard on headphones, will seem to be coming from the left ear itself. A binaural sound coming from the left can seem to be metres further out. The phase accuracy and excellent frequency response of digital recording and editing make this an ideal medium for binaural sound.

Binaural recording can make an exciting alternative for productions aimed at headphone listeners, such as a headphone tour around a historic building. Although the phase accuracy of compact cassette is not that good, the binaural effect survives, provided everything is kept digital up to the master that the cassettes are copied from. Portable CD or Minidisc players preserve the directional effects because of their inherent phase accuracy. (Small timing errors between tracks are often expressed as phase differences. Thus, a difference of 1/48 000th of a second is equivalent to a quarter of a wavelength at 1 kHz. Because of the mathematical nature of waves, a whole cycle of audio is said to have gone through 360°. Half way through a cycle is 180°. Here, complete cancellation will occur if the two sources are mixed at equal level. Cassette machines have a degree of jitter where the track relationships vary; how much is dependent on how good the tape transport is.)

Mono sound, added to the mix, remains firmly 'between the ears', and this can be used to distinguish narration from other voices. There have been a number of dramas where characters, mute from a stroke, have had 'think speech' coming from within the listener's head, while all the 'real world' action takes place outside of the head.

11.6 Surround sound

There are various systems purporting to give surround sound to the listener, and these succeed to greater or lesser degrees. In the 1970s there was a brief burst of popularity of quadrophonic sound (Figure 11.1), but this gave little regard to the actual theory of how we hear and suffered from the fact there were only two-channel delivery systems available. This meant that some form of compatible encoding matrix system had to be used. In reality there were many systems, which led to the public being confused. The need to have four full-sized loudspeakers in the room also conflicted with people's desire for space and with their interior design.

As quad began to fail commercially, there was a separate development in the cinema. Dolby Laboratories developed a way of not only dramatically improving the sound quality on ordinary optical film sound tracks but also of shoehorning in stereo and surround at the same time. They modified one of the existing quadrophonic matrix systems and reallocated the sound channels. Quadrophonic audio had used the seemingly obvious method of having front left and right speakers combined with back left and right speakers. Instead, Dolby Laboratories used the encoded channels for a centre speaker

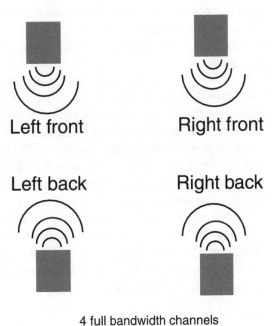

Left front   Right front

Left back   Right back

4 full bandwidth channels

Figure 11.1 Quadrophonic speaker layout

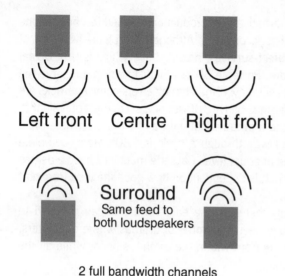

Left front   Centre   Right front

Surround
Same feed to
both loudspeakers

2 full bandwidth channels

Figure 11.2 Dolby Stereo speaker layout

and the other channel for a diffuse rear feed that was fed with surround information to give depth to the sound. The centre speaker is there to harden the centre of the sound of the centre sound image where the dialogue is put (Figure 11.2). This means that someone at the side of the audience does not feel that the dialogue is 'over there'. This still allows directional effects and even voices to come from either side or behind you. It is not good at making sounds come from beside you.

Dolby Stereo educated audiences to enjoy the surround effects on commercial films. At the same time, video recorders were becoming ubiquitous, and these increasingly had stereo playback facilities when plugged into the home viewer's hi-fi. The sound track that was available already had the surround information encoded for the cinema, and so surround sound or 'Home Theater' decoders became increasingly popular.

You need a Dolby Surround encoder to obtain surround properly. However, without this you can still get impressive results by listening to the output of your stereo audio editor while adding extra effects like thunder, rain or reverberation in mono out-of-phase. This is 'heard' by the Dolby Surround decoder as surround information. Because the system cannot reproduce every positional effect, even professionals using fully featured surround encoders will listen, as they mix, to loudspeakers fed by a surround encoder in tandem with a decoders so as to have some idea of the end effect.

Film sound tracks progressed, and new methods of encoding sound became available. At the same time DVD (Digital Versatile Disc) became available. This has the capability of handling much more data than a CD, so there is room not only for pictures but also for extra audio tracks. The standard that has been agreed upon is based on the Dolby Stereo concept, but with discrete channels to six loudspeakers. Before, the centre channel and a single surround channel were derived from a two-channel feed; now, the centre channel has it own feed and there are two rear surround feeds. This gives much stronger directional feed, although there are still positions that cannot be reproduced. Again, you should listen through a proper playback system to judge what you are achieving. Additionally, a sixth low-bandwidth channel is used to feed a subwoofer directly, instead of the feed being derived from the others (Figure 11.3). This is intended for low frequency effects such as explosions, earthquakes and thunder.

When presenting the finished material on multitrack tape, the assignment supported by the EBU and SMPTE is the following:

Ch 1   Left
Ch 2   Right

Ch 3   Centre

Ch 4   Low frequency effects

Ch 5   Left surround (or mono surround 3 dB down)

Ch 6   Right surround (or mono surround 3 dB down)

Tracks 7 and 8 can be used for Left and Right of a stereo mix.

For classical music a different system, called Ambisonics, can be used. This uses coincident microphones, and a control box effectively produces three figure-of-eight microphones, pointing forward, sideways and upwards, along with an omnidirectional output.

In practice the upward output is often omitted, as most playback systems cannot make use of height information because this needs a ceiling loudspeaker. However, full-scale 'Periphonics' can produce spectacular results.

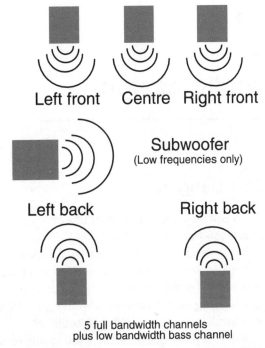

Figure 11.3 DVD 5.1 channel stereo speaker layout

11.7 Multi-layer mixing

Dramatized features sometimes need to create a sound for which there is no recording, such as the Biblical story of the Fall of Jericho. Here, building up the sound will require a great number of tracks. A large army can be simulated by sound FX recordings of suitably selected crowds, added on top of each other. If necessary, the same crowd can have different sections mixed with itself to make it sound larger. Various screams and shouts have to be added and panned appropriately, plus the trumpets that bring down the walls.

Before the arrival of digital multitrack editors this sort of thing was difficult to build up, because of the limited number of play-in machines, and often required multi-generation dubs, not to mention enormous tape loops precariously run round the room. *Cool Edit* can loop wave files so even short sections can be extended indefinitely.

Such an 'epic' can be made separately and inserted into your program, either by mixing it down and inserting it as a stereo wave file or, if you wish, totally integrating it with what precedes and follows it by appending its session (File/Append session) to your main programme session. The Append process will put the tracks on new tracks at the beginning of your programme session (to keep the Pan and Volume settings for the tracks). However, these tracks can then be grouped together so that they can be moved as a single item within the production.

12

Audio design

12.1 Dangers

Audio editors come with a multitude of 'toys' that you may never need – until that day when they rescue you from certain disaster!

Remember that just because a toy is available, you do not HAVE to use it. There is no substitute for clear, interesting, relevant speech, or good music, well performed. With documentary, have faith in the power of the spoken word. One of the more depressing of my activities as a studio manager was working with producers who did not. They would ruin exciting speech by blurring it with reverberation or obscuring it with sound effects.

All the facilities available allow you to ruin recordings very easily. Audio and music technology is full of controls that set in one direction do nothing, and set in the other make everything sound dreadful. Somewhere in between, if you are lucky, is a setting where something magical happens and a real improvement is obtained. When trying to find this setting, always refresh your ears regularly by checking the sound of the recording with nothing done to it.

12.2 Normalization

The first of these toys is normalization, which is so massively useful that you are likely to use it constantly.

One of the very first things that is done when making a recording is to 'take level'. The recording machine is adjusted to give an adequate recording level with 'a little in hand' to guard against unexpected increases in volume. However, that little bit in hand will vary, as will the volume of the speaker or performer.

Most recordists know that people are, in general, 6 dB louder on a take than when they give level – except for those occasions when they are quieter! The result is that, even for those most meticulous with levels, different inserts into a programme will vary in level and loudness. While level and loudness are connected, they are not the same thing.

Figures 12.1–12.3 show the same piece of audio but at different loudness. Figure 12.1 illustrates audio that is at a lower level (6 dB) than the other two. Figures 12.2 and 12.3 are at the same level, but Figure 12.3 sounds much louder. When we talk about level we actually mean the 'peak' level – we talk about audio 'peaking' on the meter. On average speech or music, most of the time the instantaneous level is much less.

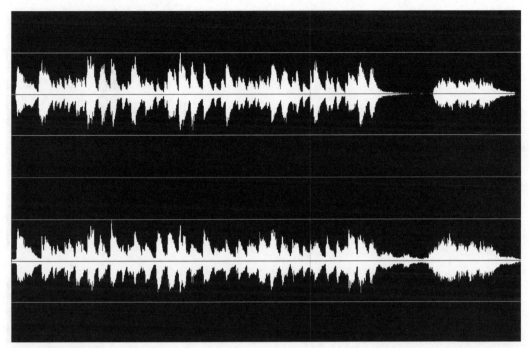

Figure 12.1 Audio at −6 dB

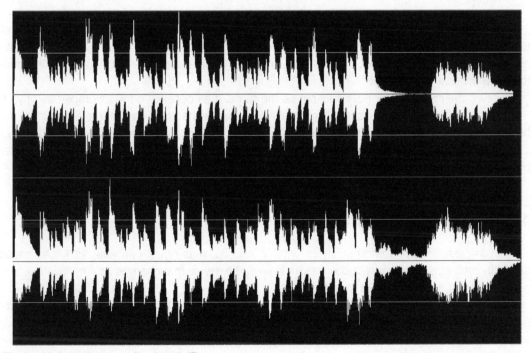

Figure 12.2 Audio normalized to 0 dB

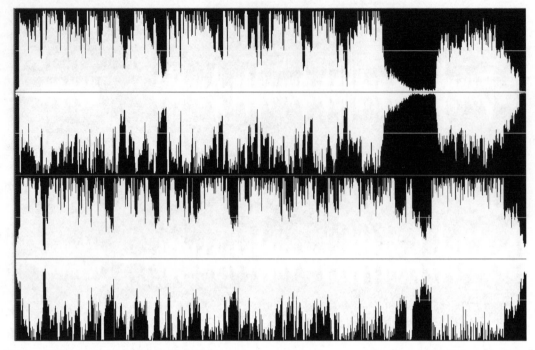

Figure 12.3 Audio compressed to 0 dB.

Audio is stored within the computer as a series of numbers, so it is very easy for the computer to multiply those numbers by a factor so that the highest number in your audio signal is set to the highest value that can be stored. If this is done to all your audio inserts when you start, it will ensure that your levels are consistent.

However, their relative loudness may not be the same. It all depends on what proportion of your recording is at the higher levels. As a generalization, while the peak level is represented by the highest point reached on the display, the loudness is equivalent to the area of the audio display.

Figure 12.3, although showing exactly the same peak level as Figure 12.2, is much louder because it has been compressed.

12.3 Compression/limiting

Audio compression varies enormously in what it can do. It is not a panacea, as it can cause audibility problems. Essentially, a compressor is an audio device that changes its amplification, and hence the level, from instant to instant.

You have control over how fast peak level is reduced (attack time), and you have separate control over how fast the level is restored after the peak (recovery time). You also have control of how much amplification is applied before compression, by how much the level is controlled, and at what level the compression starts.

The simplest compressor that most people meet is the 'automatic volume control' (AVC) used by recording devices ranging from Minidisc and video recorders to telephone answering machines. A professional recorder should give the user the option of bypassing the AVC, to enable manual control of the level.

The automatic volume control has both a slow attack and a slow recovery time. Often it has just two settings – speech and music – with the music setting having even slower attack and recovery times. Even so, modern devices found on such things as Minidisc recorders are often relatively unobtrusive.

The normal advice with analogue recording is not to use AVC where the resulting recording is going to be edited. This is because the background noise goes up and down at the same time, and any sudden change at an edit will be very obvious. With digital recording and editing the arguments are more evenly balanced, especially when digital editing is being used.

Analogue recorders overload 'gracefully', and the increase in distortion is relatively acceptable for short durations. Digital recorders are not so forgiving. They record audio as a string of numbers, and have a maximum value of number that they can record. This corresponds to peak level on the meter.

The compact disc standard, also used by Minidisc and DAT, is 16-bit, which allows 65 536 discrete levels occupying the numbers $+32\,768$ through 0 to $-32\,767$. Any attempt to record audio of a higher level than this will cause a string of numbers set at the same peak value. When played back, this produces very nasty clicks, thumps or grating distortion.

When recording items like vox pops, by the very nature of an in-the-street interview there is likely to be a very high background noise. This has always been a classic case where switching off any Automatic Volume Control was regarded as essential, because of the resulting traffic noise bumps on edits. However, the likelihood of short- or medium-term overload is quite considerable. On the other hand, sudden short, very high-level sounds such as exhaust backfires and gunshots can 'duck' the AVC to inaudibility for several seconds.

Digital editors make matching levels at edits the work of seconds. The danger of overload is such that this correction may be preferable to risking losing a recording owing to digital overload. In reality, it is down to users to decide how good their recording machine's AVC is compared with how well it copes with overloads. They may prefer deliberately to under-record, giving 'headroom'.

Most portable recorders have some form of overload limiting, which may allow you to return to base with 'usable' recordings, but it is better to return with good recordings. In the end, a saving grace is that it is the nature of these environments that the background noise is quite high and will drown any extra noise from deliberately recording with a high headroom.

Compression may make the audio louder, but it also brings up the background noise rather more than the foreground. This can mean that noise that was not a problem now becomes one. It will also bring up the reflections that are the natural reverberation of a room. This can change an open, but OK, piece of actuality into something that sounds as if the microphone had been lying on the floor pointing the wrong way.

If you are creating material for broadcast use, you should also be aware that virtually all radio stations compress their output at the feed to the transmitter. While the transmitter processors used are quite sophisticated, they will be adding compression to any you have used. This can mean that something that sounds only just acceptable on your office headphones may sound like an acoustic slum on air.

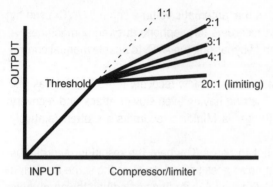

Figure 12.4 Compression ratios

There are standard options on any compressor that are emulated in any compression software. The compressor first amplifies the sound and then dynamically reduces the level (compression software is often found in the menu labelled as 'Dynamics'). Compressors can be thought of as 'electronic faders' that are controlled by the level of the audio at their input. With no input or very quiet inputs they have a fixed amplification (gain), usually controlled by separate input and output gain controls. Once the input level reaches a threshold level, their amplification reduces as the input is further increased. The output level still increases, but only as a set fraction of the increase in input level. This fraction is called the compression ratio (Figures 12.4, 12.5).

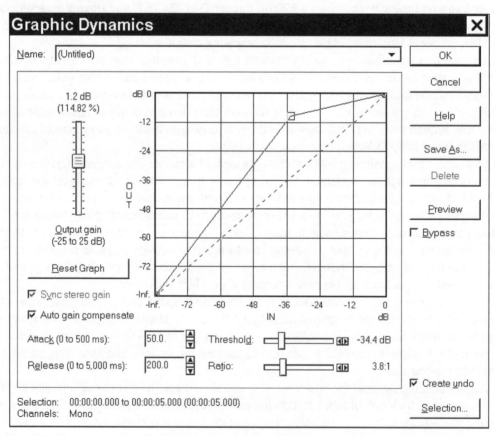

Figure 12.5 *Sound Forge* graphical dynamics display, which allows you to draw the graph of what you want

A compression ratio of 2 : 1 means that once the audio is above the threshold level then the output increases by 1 dB for every 2 dB increase in input level. A ratio of 3 : 1 means that once the audio is above the threshold level then the output increases by 1 dB for every 3 dB increase in input.

Compression ratios of 10 : 1 or greater give very little change of output for large changes of input. These settings are described as limiting settings, and the device is said to be acting as a limiter.

Hard limiting

Hard limiting can be useful for getting better level out of a recording that is very 'peaky'. Some people have quiet voices with certain syllables being unexpectedly loud. Hard limiting can chop these peaks off with little audible effect on the final recording, except that it is now louder. *Cool Edit Pro* has a separate effects transform for this.

Threshold

The Threshold control adjusts at what level the compression or limiting starts to take place. By the very nature of limiting, the threshold should be set near to the peak level required. With compression, it should be set lower; 8 dB below peak is a good starting place.

Attack

This is the control for adjusting how quickly the gain is reduced when a high level signal is encountered. If this is set to be very short, then the natural character of sounds is softened and all the 'edge' is taken out of them. It can be counterproductive, as this tends to make things sound quieter. Set too slow an attack time, and the compression is ineffective. A good place to start when rehearsing the effect of compression is 50 ms. With musical instruments much of their character is perceived through their starting transients, and too much compression with a poor choice of attack time can make them sound wrong. On the other hand, a compressor set with a relatively slow attack time can actually improve the sound of a poor bass drum by giving it an artificial attack that the original soggy sound lacks.

Release/recovery

These are alternative names for the control that adjusts how fast the gain is restored once the signal is reduced. Set it too short and the gain recovers between syllables, giving an audible 'pumping' sound that is usually not wanted. This makes speech very breathy. A good starting point is 500 ms to 1 s.

Input

This adjusts the amplification before the input to the compressor/limiter part of the circuit. With software graphical display controls this may at first appear not to be present, but it is represented by the slope of the initial part of the graph.

113

Output

This adjusts the amplification after the compressor/limiter part of the circuit. Because it occurs after compression has taken place, it also appears to change the threshold level on the output. Software compressors often have the option automatically to compensate for any output level reduction due to the dynamic gain reduction by resetting the output gain.

In/out, bypass

This provides a quick way of taking a device out of the circuit for comparison purposes.

Link/stereo

Two channels are linked together for stereo so that they always have the same gain despite any differences in levels between the left and right channels. If this is not done, you may get violent image swinging on the centre sounds such as vocals.

12.4 Expanders and gates

Cool Edit's dynamics can also act as expanders and gates (Figure 12.6). These work in a very similar way to compressor/limiters, but have the reverse effect. With high-level inputs they have a fixed maximum gain, and once the input level decreases to a threshold level then their gain decreases as the input is further decreased. When used as expanders the output levels still decrease, but only as a set fraction of the decrease in input level. This fraction is called the expansion ratio.

A expansion ratio of 2 : 1 means that once the audio is below the threshold level, then the output decreases by 2 dB for every 1 dB decrease in input level. A ratio of 3 : 1 means that the output decreases by 3 dB for every 1 dB decrease in input. Expansion ratios of 10 : 1 or greater give a very large change of output for little changes of input, and the device is said to be acting as a gate.

Unlike compressors, which were originally developed for engineering purposes, the expander is an artistic device and must be adjusted by ear. Changing its various parameters can dramatically change the sound of the sources. Its obvious use is for reducing the amount of audible spill. This is usually restricted by the fact that too much gating or expansion will change the sound of the primary instrument. However, this very fact has led to

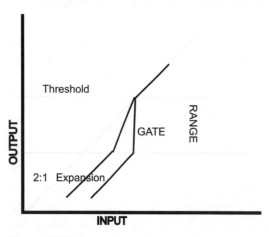

Figure 12.6 Expander and gate actions

gates and expanders being used deliberately to modify the sounds of instruments – particularly drums. Their use is almost entirely with multitrack music; bass drums are almost invariably gated.

Expanders and gates are bad news for speech. They change the background noise, and the ear tends to latch on to this rather than listening to what is said. While I can imagine circumstances where they might help a speech recording, I have not yet met any.

Dynamics transform

Cool Edit's Dynamic Effects transform is a very flexible implementation of dynamics control. It provides you with alternative ways of inputting what you want, along with a profusion of presets. A tour through these will educate your ears as to what can be done. There are four tabbed pages within the dialog box; 'Graphic', 'Traditional', 'Attack/Release' and 'Band Limiting'. The first two tabs give you alternative ways of entering what you want. All four tabs allow you access to the presets.

With the graphical tab (Figure 12.7) you can 'draw' the dynamics that you want. Click on the line to give you a 'handle', and then move it where you want. It is usually quicker to start from a preset

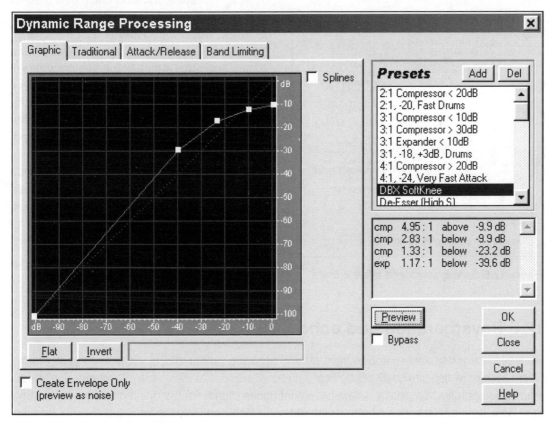

Figure 12.7 Dynamics transform – graphical tab

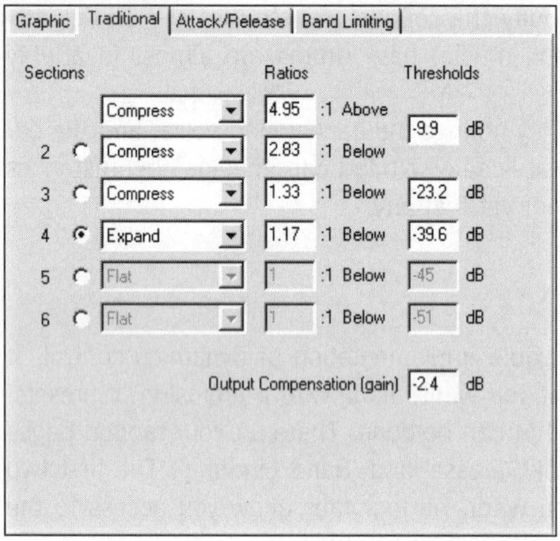

Figure 12.8 Dynamics transform – traditional tabbed page

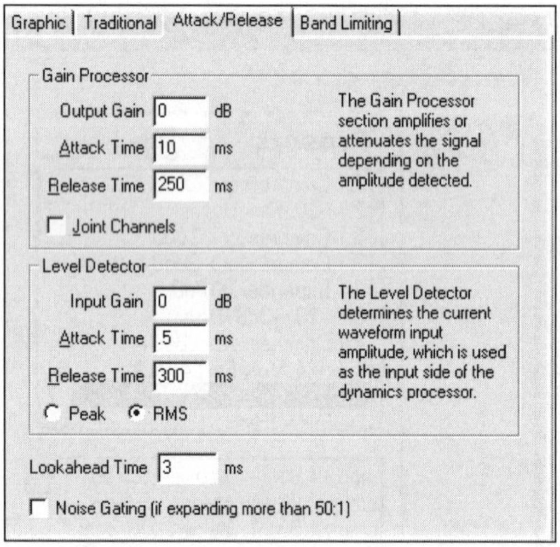

Figure 12.9 Setting attack and release times

and adjust the settings if required. Illustrated is a 'soft knee' compression of the sort associated with the dbx company.

Figure 12.8 shows the same setting in the 'Traditional' tabbed page, with each section labelled as text. Output compensation allows you to boost the level after compression, if this is needed.

Figure 12.9 is where you can set the attack and release times. How you set these times will substantially affect the sound of an instrument, as it will modify its start and finish transients. A very quick recovery time with compression will bring up background noise and increase the apparent amount of reverberation.

Figure 12.10, 'Band Limiting', allows you to restrict the frequency range at which limiting occurs. The setting shown is one that is used in one of the 'De-essing' presets. De-essers do what you would think; they reduce the level of sibilants ('S' sounds) in a vocalist's voice. You should apply this after any full frequency range compression, as otherwise the compression can restore the sibilance!

Cool Edit lacks a full-scale multiband limiter of the sort often used to make CDs and broadcasts sound louder. They split the audio into bands and separately compress each band. It's possible to simulate one using the frequency band splitting option in the multitrack editor, but at the expense of a lot of processing. DirectX multiband limiters can be bought, but good ones are expensive.

12.5 Reverberation and echo

Reverberation (often abbreviated to 'reverb') and echo are similar effects. Echo is a simpler form, where each reflection can be distinctly heard (Echo . . . Echo . . . Echo . . . Echo). Reverberation has so many different reflections that it is heard as a continuous sound. For reasons more owing to tradition than logic, broadcasters often refer to reverberation as 'echo' and use the term 'flutter echo' for echo itself.

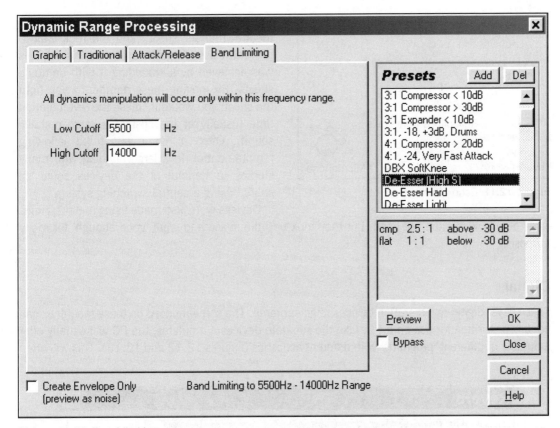

Figure 12.10 Band limiting

Reverberation time

The amount of reverberation is usually quoted in seconds of reverberation time. Reverberation time is defined as the time it takes for a pulse of sound (such as a gun shot) to reduce (or decay) by 60 dB. This ties in well with how the ear perceives the length of the reverberation.

Devices

Reverb is the first and oldest of studio special effects. Originally it was created by feeding a loudspeaker in a bare room and picking up the sound with a microphone (Figure 12.11). Although this worked, there was no way of changing the style of reverberation. It was also prone to extraneous noises ranging from traffic and hammering to underground trains or even telephones ringing inside the room.

The next development was the 'echo plate'. A large sheet of metal (the plate), about 2 m × 1 m, was suspended in a box. There was a device to vibrate the plate with audio and two pickups (left and right for stereo) to receive the reverberated sound.

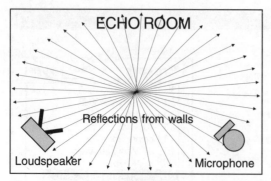

Figure 12.11 Echo room set-up

This worked on a similar principle to a theatre 'thunder sheet', and the trick was to prevent it sounding like a thunder sheet! This was achieved by surrounding it with damping sheets. By moving these dampers away from the sheet, you would increase the reverberation time (usually at the cost of a more metallic sound). Often a motor was fitted, allowing remote control. In a large building with many studios, a handful of the devices could be shared using a central switching system.

Devices were also made using metal springs. Their quality was often not good, but they took up little space and were 'good enough' for many purposes.

Digital

These days, digital reverberation devices reign supreme. The software used on these machines can equally be written for use on a PC. Like the separate devices it emulates, the PC will usually offer a number of different 'programs' with different acoustics (Figures 12.12 and 12.13). This will often

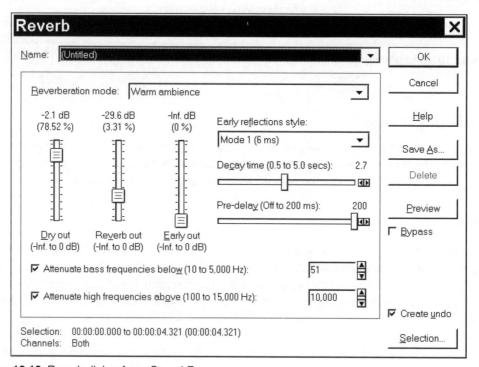

Figure 12.12 Reverb dialog from *Sound Forge*

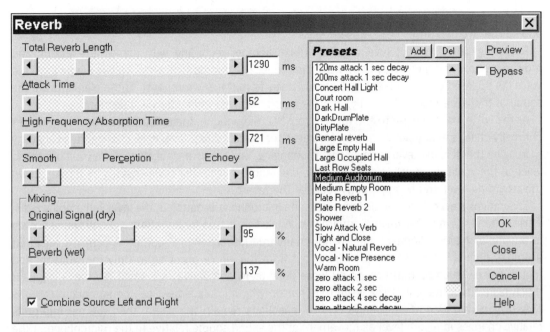

Figure 12.13 Reverb dialog from *Cool Edit Pro*, showing how presets can be added to existing ones

include emulation of the old plates and room, but also include concert hall acoustics, along with special effects that could not exist in 'real life'.

Reverberation is a complex area, as it triggers various complex subconscious cues in the brain. There are two main pieces of information we get from how 'echoey' a recording of someone speaking is, and they both interact but need separately to be controlled. These are the size of the room, and how far away from the microphone the voice is. Move the voice further from the mic and the more echoey the recording becomes. If the microphone is held and the person walks into a larger room, the more echoey it becomes. However, we are also adept at distinguishing between different types of room. We can hear the difference between an underground car park and a concert hall, a hallway and living room. How do we do this?

Let us imagine we are recording in the centre of a large, open acoustic, suspended in the air, as well as centred over the floor. Now make a sharp, impulsive noise like a single handclap, while making a recording. When we examine the waveform of the recording, we will see that there is a moment of silence between the handclap and the start of the reverberation. This is because reverberation is caused by sound reflecting off surfaces such as walls, floors and ceilings. Sound does not travel instantaneously, and its actual speed varies with the temperature and humidity of the air (this is one of the reasons why the acoustic of a hall can change so dramatically once the audience arrives, even if the seats have been heavily upholstered to emulate the absorption of a person sitting in them). Sound travels at roughly 330 m/s – about 1000 ft/s. This means that if the nearest reflecting surface is 50 feet away then you will not hear that first reflection for 100/1000th of a second (100 ms; 50 ms there and 50 ms back). It is the size of this delay that is our major cue for perceiving how large a room a recording was made in. An unfurnished living room may be as reverberant as a cathedral, but it will have short delay on the reflections from the walls (10–20 ms) and so the two can never be confused.

Returning safely to floor level, we can make another recording with someone speaking and moving towards and then away from the microphone. The microphone picks up two elements of the voice (as opposed to the ambient background noise from the environment). These elements are the sound of the voice received directly, and the sound received indirectly via reflections.

As the voice moves nearer to and further from the microphone, the amount of indirect sound does not really change in any practical way. However, the voice's direct level does change. It is very easy to imagine that it is the reverberation that is changing, and this is because a recordist will be setting level on the voice. As it gets closer it gets louder, and the recordist compensates by turning down the recording level and so apparently reduces the reverb.

Theoretically, every time the distance from the microphone is doubled, the level will drop by 6 dB. This is often known as the inverse square law. While this is true for an infinitely small sound source in an infinitely large volume of air, it is not totally true for real sound sources. A person's voice, for example, comes not only from the mouth but from the whole chest area. Only the sibilants and mouth noises (such as those made by loose false teeth and saliva) are small, so they disappear much faster than the rest of the voice (one reason why mic placement can be so critical).

As a generalization, if the delays on the reflections remain the same, then the ear interprets relative changes in reverb level as movement of the sound source relative to the microphone. If the delays change then the ear interprets this as the mic moving with the voice into a different acoustic.

The reason why reflections die away is that only a fraction of the energy is returned each time. Hard stone surfaces reflect well; soft furnishings and carpets do not. Such things as good quality wallpaper on a hard surface will absorb high frequencies, but will reflect low frequencies as well as the bare hard surface does.

Another reverberation parameter is its smoothness. Lots of flat surfaces give a lumpy, hard quality of reverberation. A hall full of curved surfaces, heavily decorated with carvings, will give a much smoother sound. This does much to explain why many nineteenth-century concert halls sound so much better than their mid-twentieth-century replacements, built when clean, flat surfaces were fashionable. So much more is now known about the design of halls that there is little excuse for getting it wrong in the twenty-first century.

When creating an artificial acoustic, all these mistakes are yours to make! As always with audio processing, it is very easy to make a nasty 'science fiction' sound but so much more difficult to create something magical and beautiful. The effects transforms within all good digital audio editors are provided with presets and it saves time to use them, albeit modifying them slightly. When you find a setting that you really like, you can save it as a new preset.

Summarizing, a simple minimalist external reverberation device would include options such as:

1 *Predelay*. This controls how much the sound is delayed before the simulated reflections are heard, and normally varies from 0 to 200 ms. As sound travels at approximately 1 ft/ms (1/1000th of a second), a predelay of 200 ms (1/5th second) will emulate a large room with the walls 100 ft (*ca.* 30 m) away (200 ft there and back).

2 *Decay*. This controls the reverberation time. This is the apparent 'hardness' or 'softness' of a room. A room with bare, shiny walls will have a long reverberation time, whereas a room with carpets, curtains and furniture will have a short reverb time.

3 *HF rolloff*. This is an extra parameter that shortens the reverb time for the high frequencies only, making the 'room' seem more absorbent and well furnished.

4 *Input and output*. This adjusts the input or output level.

5 *Mix*. This controls a mix between reverberation and the direct sound you are starting with. Sometimes there are separate controls labelled 'dry' and 'wet'. As a dry acoustic is one without much reverberation, so pure reverberation is thought of as a wet signal.

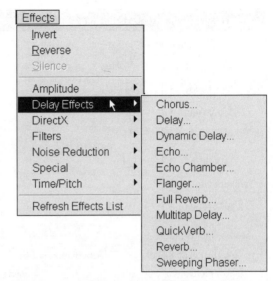

Figure 12.14 Delay effects submenu

Let us now take a closer look at how *Cool Edit Pro* implements this. Reverberation and echo comes under the submenu of delay effects, and the choice is potentially overwhelming! (Figure 12.14).

Chorus

Chorus (Figure 12.15) is an effect used a great deal in pop music. By producing randomly delayed electronic copies of a track, it can make one singer sound like several or many. It can also be used to thicken musical tracks and other sounds. The random delays inherently cause a random pitch change. This can sound very nasty, especially on spill behind a vocal.

Outside of music, chorus is not particularly useful. However, the various options provided by chorus can produce interesting stylized acoustic effects, not dissimilar to adding echo. The variable delay pitch variation can help produce effects suitable for science fiction. As well as the pitch changing caused by the variable delay, the vibrato settings introduce random amplitude changes.

Rather than me trying to describe all the effects, try changing them yourself using the demo version of *Cool Edit* on the CD supplied with this book. Using a recording of a single voice, give yourself a tour of the preset effects using the preview button. There will be sounds there that you will find useful one day.

Delay

Delay (Figure 12.16) is the equivalent of tape flutter echo where the output of the playback head was fed back to the record input. While not the most subtle effect, it does have its uses – emulating railway station announcements and 1960s rock 'n' roll come to mind. Again, give yourself a tour around the presets.

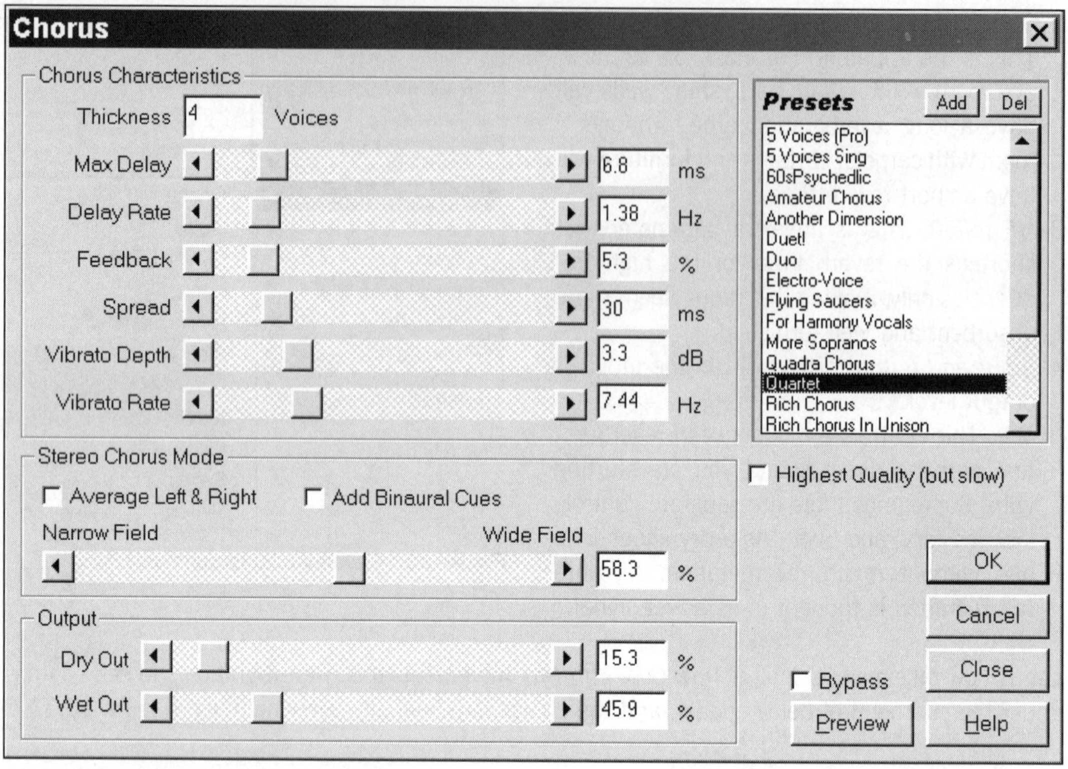

Chorus

Chorus Characteristics

Thickness 4 Voices

Max Delay ◄   ► 6.8 ms

Delay Rate ◄   ► 1.38 Hz

Feedback ◄   ► 5.3 %

Spread ◄   ► 30 ms

Vibrato Depth ◄   ► 3.3 dB

Vibrato Rate ◄   ► 7.44 Hz

Presets [Add] [Del]

5 Voices (Pro)
5 Voices Sing
60sPsychedlic
Amateur Chorus
Another Dimension
Duet!
Duo
Electro-Voice
Flying Saucers
For Harmony Vocals
More Sopranos
Quadra Chorus
Quartet
Rich Chorus
Rich Chorus In Unison

☐ Highest Quality (but slow)

Stereo Chorus Mode

☐ Average Left & Right   ☐ Add Binaural Cues

Narrow Field   Wide Field

◄   ► 58.3 %

Output

Dry Out ◄   ► 15.3 %

Wet Out ◄   ► 45.9 %

[OK] [Cancel] [Close] ☐ Bypass [Preview] [Help]

Figure 12.15 Chorus effect

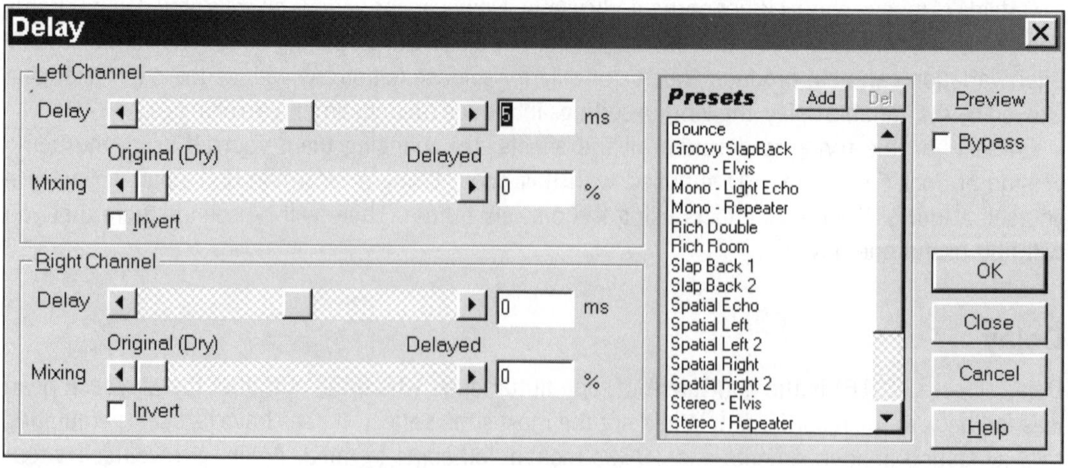

Delay

Left Channel

Delay ◄   ► 5 ms

Original (Dry)   Delayed

Mixing ◄   ► 0 %

☐ Invert

Right Channel

Delay ◄   ► 0 ms

Original (Dry)   Delayed

Mixing ◄   ► 0 %

☐ Invert

Presets [Add] [Del]

Bounce
Groovy SlapBack
mono - Elvis
Mono - Light Echo
Mono - Repeater
Rich Double
Rich Room
Slap Back 1
Slap Back 2
Spatial Echo
Spatial Left
Spatial Left 2
Spatial Right
Spatial Right 2
Stereo - Elvis
Stereo - Repeater

[Preview] ☐ Bypass [OK] [Close] [Cancel] [Help]

Figure 12.16 Delay

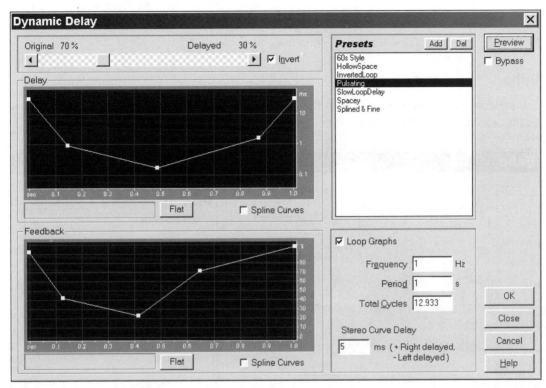

Figure 12.17 Dynamic delay

Dynamic delay

The dynamic delay effects transform allows you to flange the sound in a graphically controlled way (Figure 12.17). By varying the delay between the original and a copy, you get a comb filter effect – the 'psychedelic' sound associated with sixties pop music.

In those days the only practicable way of doing this was to feed the music to be treated to two identical tape recorders, and combine the outputs of the two playback heads. Because of small mechanical imperfections, the two delays introduced by the replay heads were very slightly different. They would vary slightly as the tension varied slightly and changed the stretch of the tape. This became known as *phasing*. It causes a notch in the frequency response, giving a drainpipe type sound. This is because the two signals cancel where half the recorded wavelength of the audio becomes equal to the delay. Depending on the delays involved, not only will the frequency cancelled-out change but there may well be a multiplicity of cancellations producing what is known as a comb filter effect.

According to legend, it was John Lennon of The Beatles who found that the delay difference could be modified subtly and controllably by putting a finger on the flange of the feed spool of one of the tape recorders. The effect became known as *flanging*. It gives the classic skying sound of the sixties. This is a comb filter effect moving up and down the sound spectrum.

Used as a real-time effect in the multitrack view, dynamic delay's parameters can be controlled by an envelope.

Echo

Echo (Figure 12.18) is very much like the delay effects transform with the additional option of adding equalization to the delayed echo. Again, some useful presets are supplied.

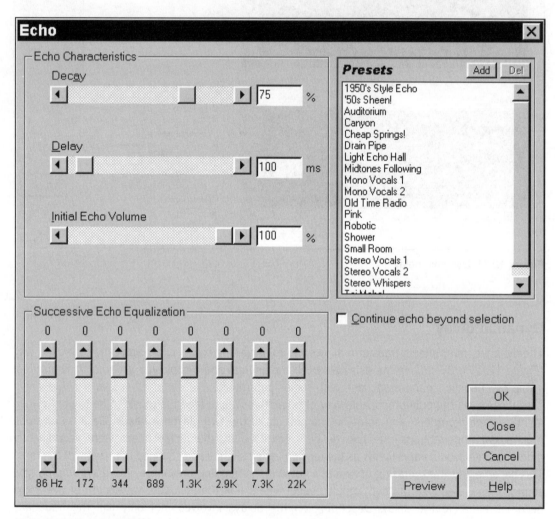

Figure 12.18 Echo effect

Echo chamber

The echo chamber approaches designing a reverberant effect in a totally different way (Figure 12.19).

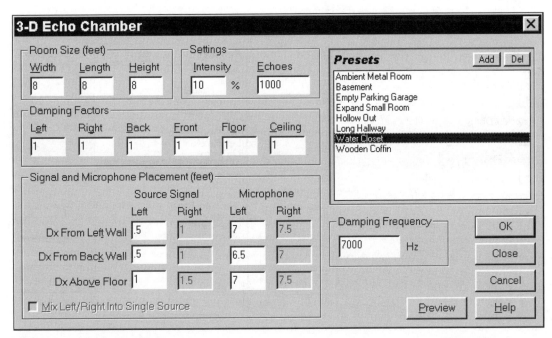

Figure 12.19 3D echo chamber

You are invited to enter the dimensions of the room you wish to simulate, and to identify where the microphone should be placed. You can also set the absorption (damping factor) of the floor, ceiling and each wall. Some useful dramatic environments are supplied as presets.

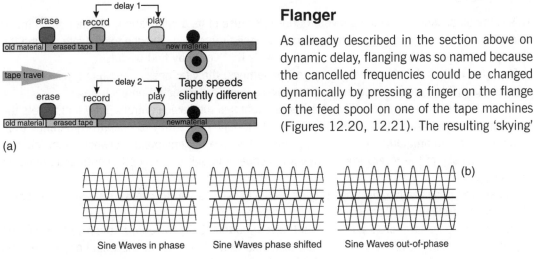

Flanger

As already described in the section above on dynamic delay, flanging was so named because the cancelled frequencies could be changed dynamically by pressing a finger on the flange of the feed spool on one of the tape machines (Figures 12.20, 12.21). The resulting 'skying'

Figure 12.20 (a) Original tape phasing. (b) Phasing/flanging. Delaying the bottom wave leads to it being out-of-phase with and therefore cancelling the top wave

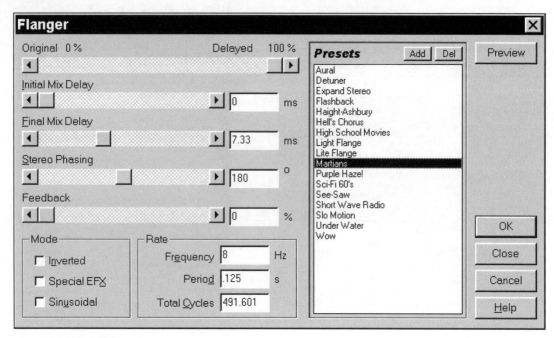

Figure 12.21 Flanger effect

effect is irretrievably associated with late 1960s pop music. Both phasing and flanging are available, and have their own useful presets.

Full reverb

This is the all-singing, all-dancing reverb option. Because of its complication, I do not recommend using this unless you have a specific need and have the time and knowledge to navigate around the options. However, as before, there are some presets that you may find useful.

There are no less than three tabbed pages for settings; the vital ones controlling the level of the original signal (dry) early reflections and the reverb (wet) are repeated on each page.

The first page, general reverb (Figure 12.22a), contains the basic controls. Total length sets the reverberation time; attack time sets the predelay (although *Cool Edit Pro* operates differently from some other reverb programs in this regard); and diffusion and perception, between them, change the characteristic of the reverb from lumpy and echoey through to a smooth, continuous build-up and decay.

The second page, early reflections (Figure 12.22b), allows you to control the early reflections, effectively controlling the room size.

The third page, coloration (Figure 12.22c), gives you the ability to equalize the reverb within the algorithm. You can move the three bands, low, mid and high, around the frequency band. The amount of coloration is adjusted by the three sliders. As with the other effects transforms, the easiest approach is to use the presets and then modify as required.

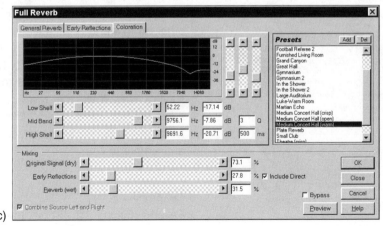

Figure 12.22 There are three tabbed pages for full reverb settings: (a) general reverb; (b) early reflections; and (c) coloration

Multitap delay

Multitap delay (Figure 12.23(a)) simulates the old tape loop delay systems (Figure 12.23(b)). The effects produced are relatively primitive, but suit certain types of music or stylized effects.

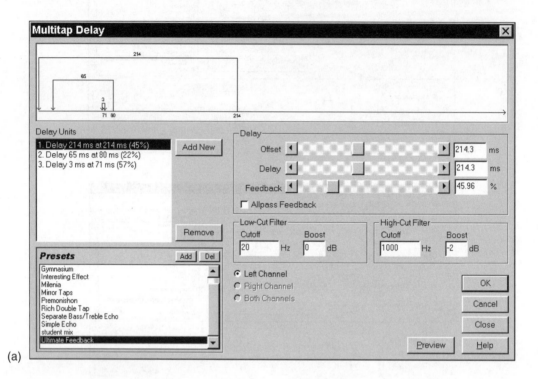

(a)

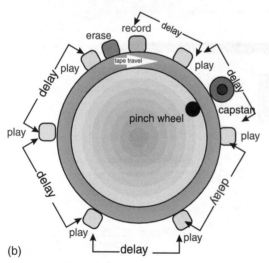

(b)

Figure 12.23 (a) Multitap tape delay; (b) the old tape loop delay system.

Quickverb

Quickverb (Figure 12.24) is a very basic reverb transform. It uses a simpler algorithm than the other reverbs and is intended for use with real-time processing, where the reverb does not have to be of the highest quality but allows the processing required to be kept to a minimum. Having said that, the reverb is more than adequate for most tasks. Like the 'reverb' transform about to be described, it is flexible and simple to operate.

Reverb

Reverb (Figure 12.25) is the most useful general-purpose reverberation effects transform. It is flexible, but still simple to operate.

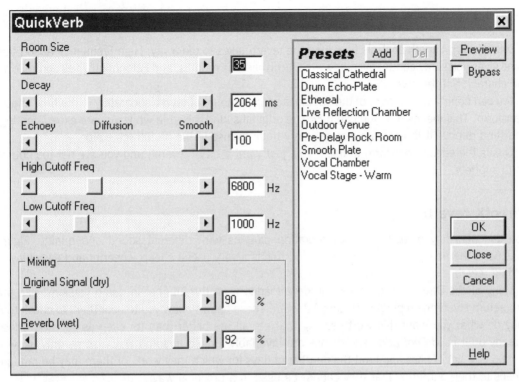

Figure 12.24 Quickverb effect

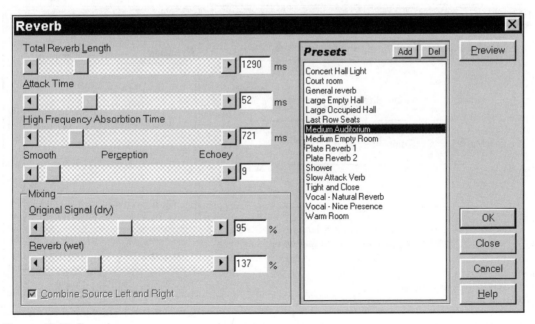

Figure 12.25 Reverb

As well as a sensible range of presets, the reverb effects transform has the basic logical controls.

Total Reverb Length controls how long the reverb takes to die away; High Frequency Absorption Time gives you some control over 'how well furnished' the room is; and Perception gives control over the character of the reverb.

You can control the amount of the original and the amount of reverb separately in the final effects transform. This means that you can have the original sound mixed in while you are experimenting, and then remove it to generate a reverb-only track for use in the multitrack editor.

This is the effects transform to use if you 'just want a bit of reverb', and you are not too critical as to subtlety.

DirectX reverb

You can also buy or download reverberation plug-ins with different sound and quality. Figure 12.26a is the free 'FreeverbX' plug-in. I find this is a very useful cheap, cheerful and easy way of adding reverb.

Many of the DirectX plug-ins can be downloaded from the Internet in demo versions, like the Arboretum Hyperprism plug-in. (Figure 12.26b). These demos allow you to hear them to check that they do what you want. It's worth saying that not all are better than (or even as good as) those already built-in to *Cool Edit*. The demos are invariably limited in some way – some only allow a certain number of uses and limit the number of days for which they work, or there may be random bleeps in their output so that they cannot be used in a practical way.

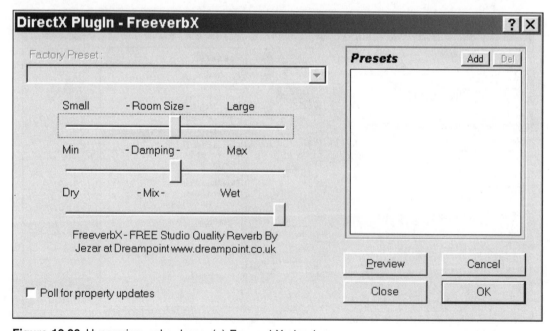

Figure 12.26 Hyperprism echo demo: (a) FreeverbX plug-in

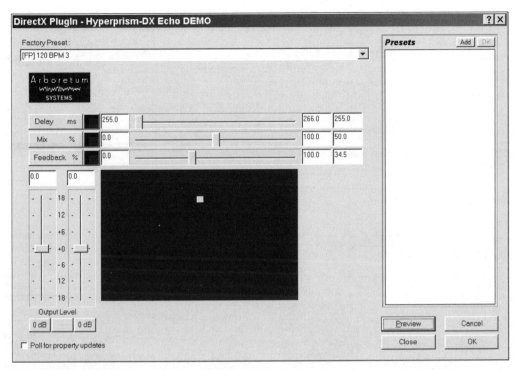

Figure 12.26 Hyperprism echo demo: (b) Hyperprism-DX Echo demo

Many plug-ins come with a whole suite of effects. These will often appear as a submenu in *Cool Edit Pro*, as is the case with Hyperprism DX (Figure 12.26c)

Sweeping phaser

The sweeping phaser (Figure 12.27) is very similar to the flanger, but with much more control over the effects available.

Again, start by using the presets and play with the settings to modify the sounds.

12.6 Equalization

Cool Edit Pro has almost an embarrassment of equalization and filters. Equalization (Figure 12.28), or EQ as it is usually known, has two functions: the first is to enhance a good recording, and the second is to rescue a bad one.

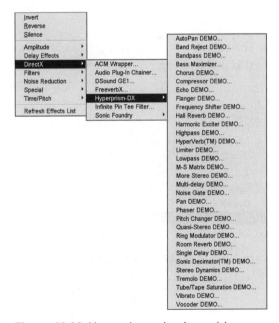

Figure 12.26 Hyperprism echo demo: (c) Submenu in *Cool Edit Pro*

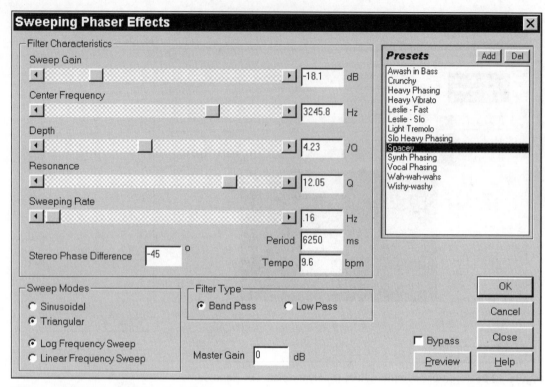

Figure 12.27 Sweeping phaser effects

The best results are always obtained by first choosing the best microphone and placing it in the best place. This is not always possible, and so, in the real world, EQ has to come to the rescue.

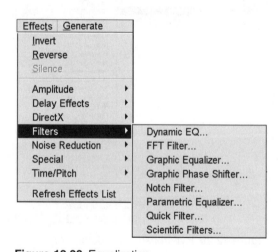

Figure 12.28 Equalization

The most common need for speech EQ is to improve its intelligibility. A presence boost of 3–6 dB and around 2.8 kHz usually makes all the difference. Here, the simple graphic equalizer effects transform will be fine.

Another common problem is strong sibilance, where 'S's are emphasized. Here, reducing the high frequency content can be advantageous (another option is to use the De-essing preset in the Amplitude/dynamics preset. This has the effect of reducing the high frequencies only when they are at high level).

Bass rumbles are usually best taken out by high-pass filters. This naming convention can be confusing at first, as it has a negative logic. A filter that passes the high frequencies is not passing the low frequencies! Therefore a high-pass filter provides a bass cut, and a low-pass filter provides a high frequency cut. This sort of filter works without changing the tonal quality of the recording.

As always, it is best to avoid the problem in the first place. If your recording environment is known to have problems with bass rumbles, such as in Central London radio studios that have underground trains running just below them, then use a microphone with a less good bass response. A moving coil mic will not reproduce the low bass well, unlike an electrostatic capacitor microphone.

Dynamic EQ

Dynamic EQ (Figure 12.29) allows you to vary the effect over the selection. There are three graphs, selected by the tabs at the top. The first tab allows you to vary the operating frequency with time, the second tab the gain, and the third tab the 'Q' or bandwidth. Three types of filter are provided; low-, band- and high-pass. The first and last terms can be confusing – a low-pass filter removes high frequencies (and passes low frequencies), while a high-pass filter does the opposite. A band-pass filter passes frequencies within its range while filtering those outside. It can be thought of as a 'middle pass' filter, although it can also be used at the extremes of the frequency range.

When used as a real-time effect in the multitrack view; gain, frequency and 'Q' can either be controlled manually or assigned to a coloured envelope control.

FFT filter

FFT stands for Fast Fourier Effects transform (Figure 12.30), and is nothing less than the techie's dream of designing a totally bespoke filter – great fun if you know what you are doing, but totally confusing if you do not! It is possibly best avoided if you did not start your career as a technician.

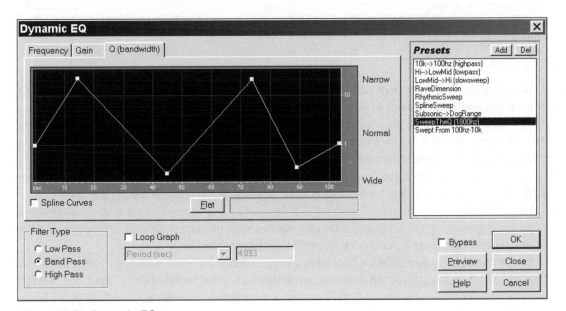

Figure 12.29 Dynamic EQ

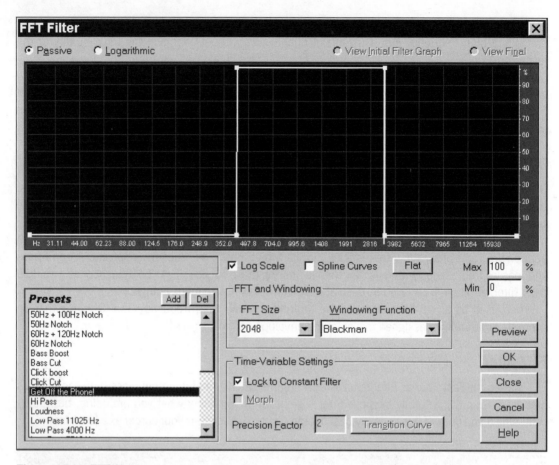

Figure 12.30 FFT filter

Designing your own filter does not need any mathematical ability, but you have to have some idea of what frequencies you want to filter. You can literally design a filter to remove frequencies that have broken through on to your recording at a gig. You do this by drawing the frequency response that you want. The line on the graph is modified just like the envelopes on the multitrack mixer – click on the line to create a 'handle' and then use the mouse to move it. If the spline curves box is ticked then a smoother curve is achieved, with the handles acting more as though they are attracting the line toward themselves rather than actually being 'fence posts' on the line.

The windowing functions are different ways of calculating the filters. There is a trade-off between accuracy in generating the curve and artefacts that can spoil the sound. Syntrillium recommend the Hamming and Blackman functions as giving the best overall results.

You can also set up two different filter settings and automatically get a transition between them. This will either be a crossfade between the two settings or, if the 'morph' box is ticked, the settings will gradually change. This means that if you have a transition between a lot of bass boost and a lot of top boost, morphing will cause the boost, in Syntrillium's words, to 'ooze' along the frequencies between one and the other. With the box unticked, the frequencies between are not affected. The precision factor controls how large the steps are in the transition; a higher number means smaller steps but more processing time.

The FFT size controls the precision of processing higher numbers, giving better resolution but slower processing time. Setting to a low value like 512 gives speed when previewing and listening to how your do-it-yourself filter performs. When you actually want to run the filter, set it to a higher value for better results – Syntrillium recommend values from 1024 to 8196 for normal use. The power-of-two values in the drop down are the only ones that will work.

For the best results, filter using 32-bit samples. If your source audio is less than this, then convert the file 32-bit to do the filtering and, when done, convert back to the lower resolution. This will produce better results than processing at lower resolutions, especially if more than one transform will be performed on the audio. This is because higher resolution mathematics will be used for the 32-bit file and more accurate results obtained.

Graphic equalizer

This effects transform emulates the graphic equalizer often found on hi-fi amplifiers. Rather than just offering the five or six sliders of the average hi-fi, it gives you a choice of 10, 20 or 30 sliders. Each covers a single frequency band. The more sliders, the better the resolution. It represents a more friendly way of designing your own frequency response, and works rather better than analogue equivalents as it does not suffer from the analogue system's component tolerances. Figure 12.31 shows a setting for a typical presence boost using 10 or 30 sliders.

Parametric equalizer

The parametric equalizer (Figure 12.32) is another techie-orientated effects transform. While there is a graphical display like the FFT filter, here the control is applied by putting in figures or by moving sliders. The sliders each side of the graph control the low and high frequencies, while the horizontal slider underneath sweep the centre frequencies of five 'middle' controls. The vertical sliders to the right control the amount of boost or cut for each filter. Most mixers that have middle control usually only have two knobs; the '±' control and the frequency sweep. More expensive mixers also have a 'Q' knob, which controls the width of the cut or boost – a very high number makes narrow notches, while a low number makes the boost or cut very wide. Technically, 'Q' is the ratio of width to centre frequency. *Cool Edit*'s parametric equalizer also has the option of constant bandwidth as measured in Hertz wherever you are in the frequency range.

Notch filter

The notch filter (Figure 12.33) is a specialist effects transform that you hope never to have to use. It allows you to set up a series of notches in the frequency response to remove discrete frequencies. Figure 12.32 shows the parametric equalizer set to filter 50 Hz and harmonics, and it comes ready set up for undertaking just this, the most common disaster recovery task: getting rid of mains hum

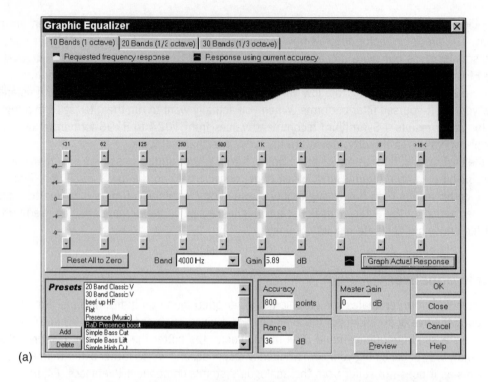

(a)

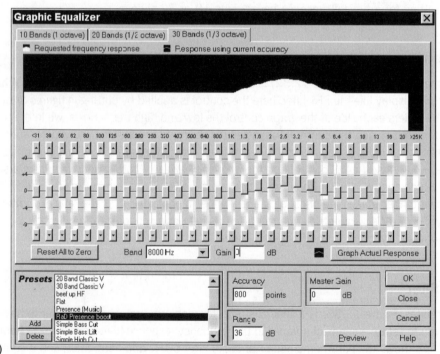

(b)

Figure 12.31 Graphic equalizers, with (a) 10 sliders; (b) 30 sliders

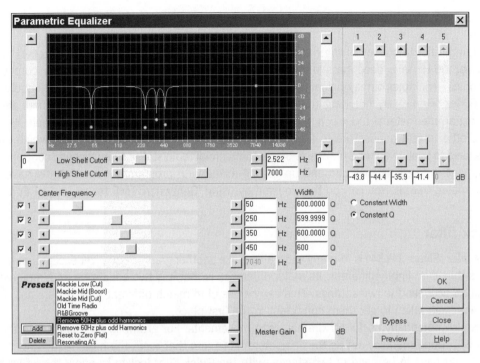

Figure 12.32 Parametric equalizer set to 50 Hz plus odd harmonics

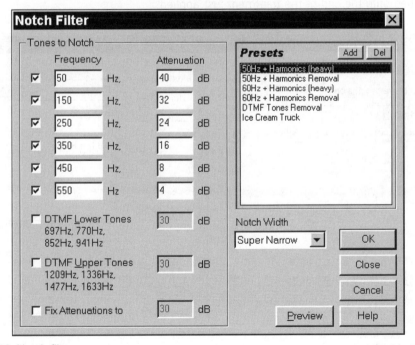

Figure 12.33 Notch filter

and buzz resulting from poorly installed equipment (theatre lighting rigs are a common source of buzzes).

In Europe the frequency of the mains power supply is 50 Hz, and in the USA it is 60 Hz. Pure 50 Hz or 60 Hz is a deep bass note. However, mains hum is rarely pure and it comes with harmonics – multiples of the original frequency – 100 Hz, 150 Hz, 200 Hz, etc. (or their equivalents for a 60 Hz original). By 'notching' them out, the hope is that they can be removed without affecting the programme material too much. Except in severe cases, this works surprisingly well.

The other preset options are to filter DTMF (Dual Tone Multi-Frequency) tones, which are the tones used to 'dial' telephone numbers. Some commercial radio stations use these to control remote equipment. Apparently they can end up getting mixed with programme material and have to be unscrambled from it. You can do this yourself by using 'Generate/DTMF tones' in *Cool Edit*.

Quick filter

Quick filter (Figure 12.34) is in many ways a little like another graphic equalizer except that the way it does things is slightly different. Its major difference is that it is two equalizers and implements a transition between the two settings. This can be useful to match different takes where not enough care has been taken to get them consistent in the first place.

It is also potentially useful for drama scene movements. For example, an orchestra in the next room will sound muffled, but as the characters move into that room the high frequencies appear. This would have to be done in two stages, with the Quick filter locked to apply the 'other room' treatment up to the point of the transition. It is then unlocked and the change made over a short section, butted onto the end of the treatment just applied.

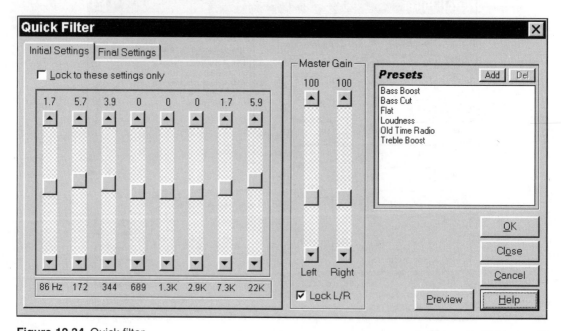

Figure 12.34 Quick filter

However, I strongly recommend not doing it this way. Rather, make a new track with the 'other room' treatment on it and use the multitrack editor to manage the change instead (see section 9.8, Transitions).

Scientific filters

The scientific filter (Figure 12.35) is another build-your-own filter kit. Analogue filters are effectively made of building blocks, each of which can only change the frequency response by a maximum of 6 dB per octave. So, if you need something sharper, then an 18 dB/8ve filter requires three building blocks. This sort of filter is known as a third-order filter.

In the analogue domain, these filters are built from large quantities of separate components. Each of these components will have manufacturing tolerances of 10 per cent, 5 per cent, 2 per cent or 1 per cent, depending on how much the designer is prepared to pay. This means that anything more than a third- or fourth-order filter is either horrendously expensive or impossible to make, because the component tolerances blur everything.

In the digital domain, these components exist as mathematical constructs. They have no manufacturing errors, and therefore are exactly the right value. This means that a digital filter can

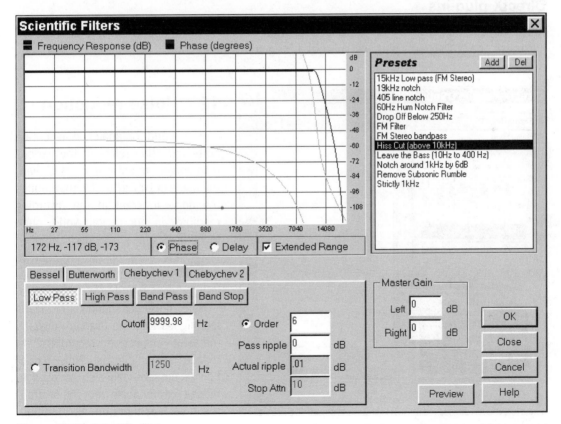

Figure 12.35 Scientific filters

have as many orders as you like, limited only by the resolution of the precision of the internal mathematics used.

Even so, you cannot get something for nothing. Heavy filtering will do things to the audio other than change its frequency response. Differential delays will be introduced along with relative phase changes, and some frequencies will tend to ring. Leaving the jargon aside, what this means is that you may achieve the filtering effect that you want, but with the penalty of the result sounding foul.

The scientific filter effects transform offers a set of standard filter options from the engineering canon. The graph shows not only the frequency response, but also the phase or delay penalties. In the end your ears will have to be the judge. If you have a tame audio engineer to hand – assuming that you are not one yourself – then you may persuade him or her to design you some useful filters optimized for your requirements and saved as presets.

A neat trick to eliminate phase shifts can be to create a batch file that runs the filter you want, then reverses the files and runs the filter again – effectively backwards. It then reverses the file back to normal. The idea is that the shifts cancel themselves out. It also means that you can use a 9th order filter twice, rather than an 18th order one once.

DirectX plug-ins

Again DirectX plug-ins can be downloaded often with specialist applications. Figure 12.36 illustrates one of Dsound's plug-ins optimized for guitar equalization.

Figure 12.36 DirectX plug-in – D Sound GE1

12.7 Noise reduction

In the world of analogue recording noise reduction refers to the various techniques, mainly associated with the name of Dr Ray Dolby, that reduce the noise inherent in the recording medium. In our digital world, such techniques are not needed except in some data reduction systems such as the NICAM system used for UK television stereo sound.

Here, noise reduction (NR) means techniques that are used to 'digitally remaster' analogue recordings to reduce their deficiencies. Noise reduction techniques can also be used to reduce natural noise from the environment, such as that from air conditioning or even distant traffic rumble.

Be warned that this is not magic; there will always be artefacts in the process. A decision always has to be made regarding whether there is an overall improvement. Be very careful when you first use these techniques, as they are very seductive. It takes a little while to sensitize yourself to the artefacts, and it is very easy to 'overcook' the treatment. As you would expect, the more noise you try to remove, the more you are prone to 'nasties' in the background.

The most common noise reduction side effect is a reduction of the apparent reverberation time as the bottoms of reverb tails are lost with the noise. In critical cases it can be necessary to add well-chosen artificial reverb at very low level, to fill in these tails. Getting a good match between the original and the artificial requires good ears.

The basic method of noise reduction is very simple. You find a short section (about 1 second) of audio that is pure noise, either during in a pause, or the lead in to the programme material. This is analysed and used by the program to build up a filter and level guide. Individual bands of frequencies are analysed in the rest of the recording, and when any frequency falls below its threshold in what is sometimes known as the fingerprint, then this is reduced in level by the set amount. The quieter the noise floor, the more effective this is.

However, trying to reduce high wide-band hiss levels can leave you with audio suitable only for a science fiction effect. The availability of NR should not be used as an excuse for sloppiness in acquisition of recordings. Don't neglect obtaining the very best quality original recording in the first place.

The classic example of this is old 78 rpm recordings. These used a much wider groove. Theoretically you could use an LP gramophone at 33 rpm to dub a 78, and then speed correct and noise reduce in the computer. However, this would sound foul. Played with the correct equipment, with purpose-designed cartridges and a correct sized stylus at the correct playing weight, the quality from a 78 can be astonishing. Remember that most of them were 'direct-cut'; they were created direct from the output of the live performance mixer. There was no intervening tape stage. Old mono LPs had wider grooves than stereo records, and a wider stylus will produce better outcomes. A turntable equipped with an elliptical stylus will produce the best results from both mono and stereo LPs.

The most effective noise reduction can be achieved just by cleaning the record. Even playing the record through, before dubbing, can clean out a lot of dirt from the groove.

A really bad pressing, with very heavy crackles, will sometimes benefit from wet playing. Literally, pour a layer of distilled water over the playing surface and play it while still wet. This can improve less noisy LPs as well, but once an LP has been wet played, it usually sounds worse once dry again and needs always to be wet played. If it is not your record, then the owner may have an opinion on this!

The subject of gramophone records reminds us of the other major form of noise reduction, namely declicking. There are a wide number of different specialist programs that do this, as well as general noise reduction. They vary regarding how good and how fast they are. If you are likely to be doing a lot of noise reduction, then it may be worth you investing in one of these specialist programs.

Some of them, like the *Sound Forge* DirectX Noise Reduction plug-in, do just that – they plug into existing programs. *Cool Edit Pro* has its own effective NR effects transforms, but can equally well use a DirectX plug-in like the *Sound Forge*.

When *Cool Edit Pro* detects a DirectX plug-in it creates a new menu option under Effects transform, which cascades submenus to allow selection of the plug-ins available (Figure 12.37).

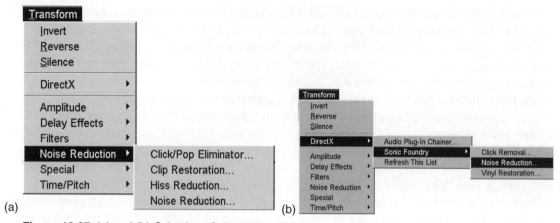

Figure 12.37 (a) and (b) Selection of plug-ins available

These can include equalization and special effects as well. Some of these plug-ins can cost more than the program that they plug into!

Figure 12.38 shows the beginning of the vocal track from a cassette-based four-track machine with its dbx noise reduction switched off. Looking closely at the track you can see that some of the noise is from tape hiss and hum, but there is also spill from the vocalist's headphones. I have selected a couple of seconds where there is no spill, only machine noise.

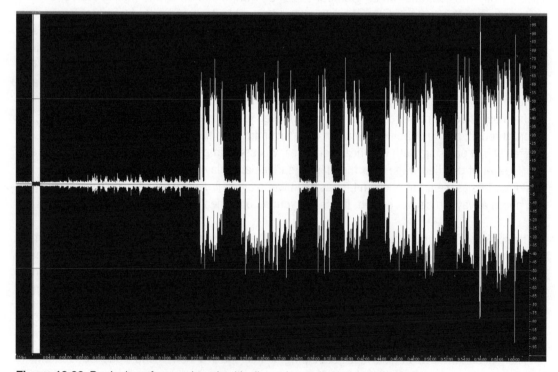

Figure 12.38 Beginning of a vocal track with dbx noise reduction switched off

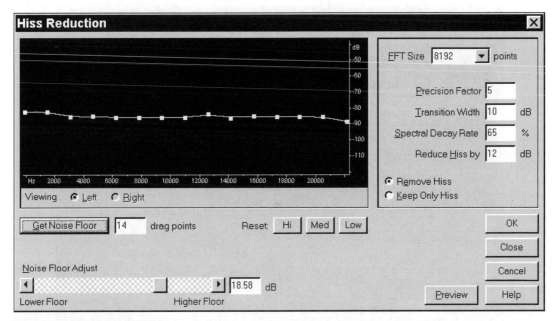

Figure 12.39 Hiss reduction

I now need to obtain a noise print as a reference for treating the whole file. Both *Cool Edit Pro* and the *Sound Forge* plug-in have similar approaches. *Cool Edit Pro* also has a hiss reduction option. This produces a noise print looking solely for hiss (Figure 12.39).

However, if we take a noise print using the noise reduction effects transform, we can see that it has picked up the hum on the output of the four-track machine and raises the threshold at the hum frequencies (Figure 12.40).

The *Sound Forge* noise reduction plug-in does the same (Figure 12.41).

Having taken the noise print, we return to *Cool Edit Pro* and select the whole wave and initiate the noise reduction process. In Figure 12.42, for clarity I used a high reduction setting available from the *Sound Forge* plug-in, so that you can clearly see that it has removed the hiss but left the headphone spill. It will depend on the material whether this amount of noise reduction sounds acceptable. The spill can be removed selectively using the Silence effects transform.

Click/Pop/Crackle Eliminator (Figure 12.43)

While *Cool Edit's* declicker is not as fast as some, it is very effective and has improved with each revision of the program. You can correct an entire selection or instantly remove a single click if one is highlighted at a high zoom level. You can use the Spectral View feature with the spectral resolution set to 256 bands and a Window Width of 40 per cent to see the clicks in a waveform. See the Spectral Display area of Options > Settings > Display to adjust these parameters. Clicks will ordinarily be visible as bright vertical bars that go all the way from the top to the bottom of the display. To satisfy yourself that only clicks rather than music were removed, you can save a copy of the original file somewhere, then Mix Paste it (overlap it) over the corrected audio with a setting of 100 per cent and Invert enabled.

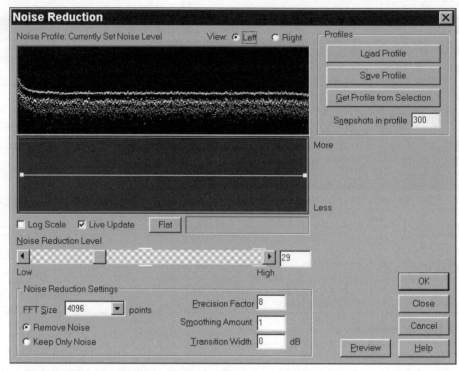

Figure 12.40 Using the noise reduction transform

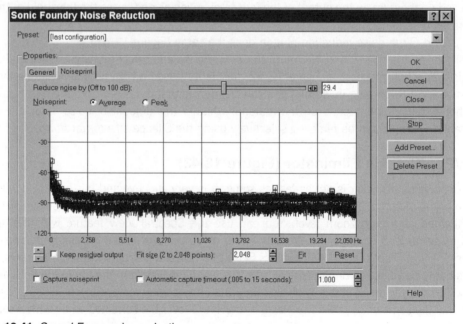

Figure 12.41 *Sound Forge* noise reduction

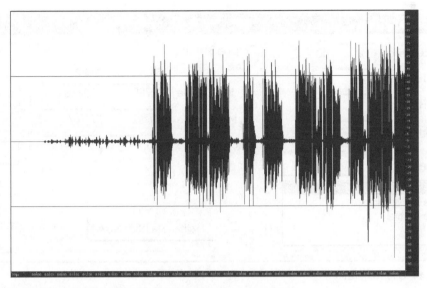

Figure 12.42 Hiss has been removed, but headphone spill remains

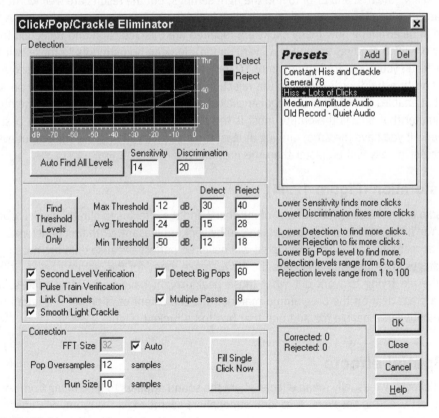

Figure 12.43 Click/pop/crackle eliminator

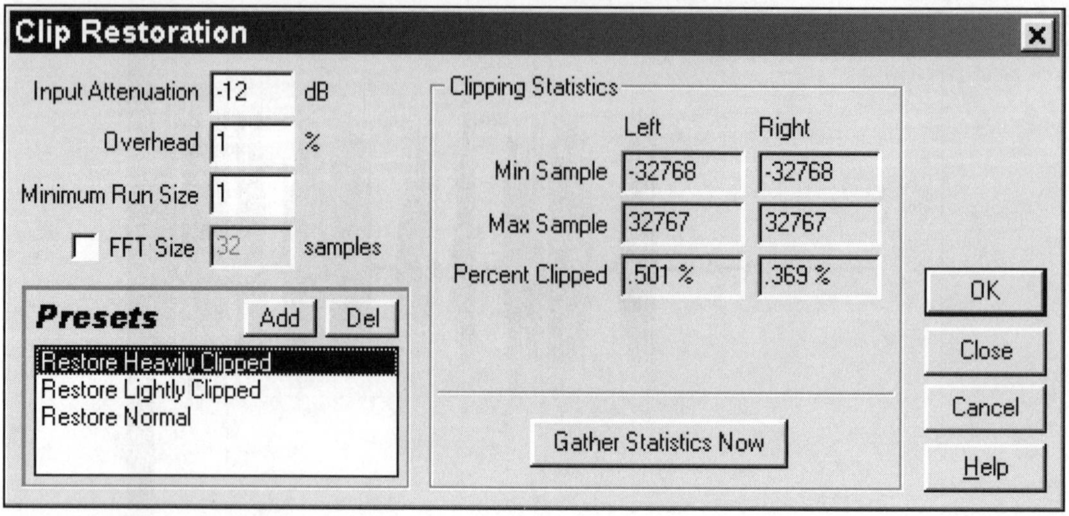

Figure 12.44 Clip restoration

It may take a little trial and error to find the right settings, but the results are well worth it – much better than searching for and replacing each click individually. The parameters that make the most difference in determining how many clicks are repaired are the Detection and Rejection thresholds (the latter of which requires second level verification). Making adjustments to these will have the greatest effects; you might try settings from 10 for a lot of correction, 50 for very little correction on the detection threshold, or 5 to 40 on the rejection threshold. Second level verification slows the process, but enables *Cool Edit* to distinguish clicks from sharp starting transients in the music. The next parameter that affects the output most is the run size. A setting of about 25 is best for high quality work. If you have the time, running at least three passes will improve the output even more. Each successive pass will be faster than the previous one.

Clip restoration (Figure 12.44)

If your audio is overloaded, you 'run out of numbers' and the effect is to clip the peaks, giving them flat tops. You'll hear this as distortion, which can be very unpleasant. With this transform, *Cool Edit* has a fist at restoring the audio, although it will never be perfect because data are missing. It does it in two stages. The first is to attenuate the file to 'make room' for the restored peaks. It then goes through the file trying to work out what those peaks might have been. How successful this is depends a great deal on the programme material and the extent of clipping, but it can make the difference between a usable file and one that has to be junked (Figure 12.45).

12.8 Spatial effects

By spatial effects, we usually mean effects where the sound appears to be coming from somewhere other than between the speakers. A general atmosphere seems to fill the room, or a spaceship flies overhead from behind you.

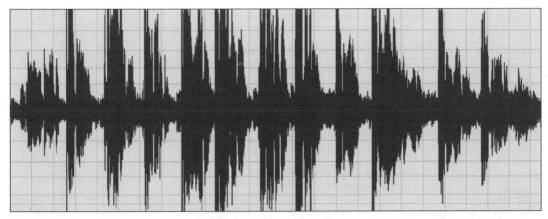

(a)

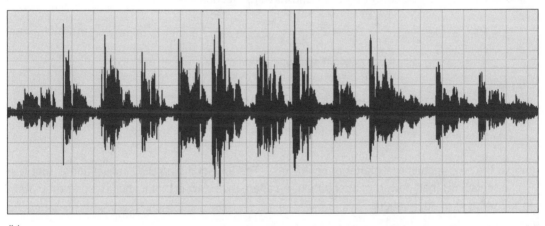

(b)

Figure 12.45 Clipped waveform (a) before processing, and (b) after processing

Techniques like Dolby stereo use extra loudspeakers and special processing to get their directional effect. DVD recordings can use five channels of audio to move sounds, but simple stereo can be beefed up, with a little simple manipulation.

Normally we think of stereo as consisting of left and right channels. The relative volume of these channels controls where a sound comes from. A more sophisticated way of thinking about stereo is as an MS signal – MS standing for 'middle' and 'side', *not* mono and stereo. The middle signal consists of a simple mix of the left and right signals in phase. The side signal is again a simple mix of the left and right signals, with the right channel phase reversed (inverted).

The middle signal is what is used as the mono signal in broadcasting. If you feed your stereo channel with a simple mono signal panned to the centre – say a presenter's voice – then there will be no side signal; the left and right channels are identical and, because the right-hand signal is inverted, they cancel themselves out.

As the mono signal is panned to one side the side signal increases until the sound is panned fully to one side. At this point, the middle and side signals are equal in level. Once you are thinking in this way, you can consider what happens when you increase the side signal so that it is larger than the middle signal. Doing this can pan the sound image outside the speakers, and how effective this will be depends on the programme material. As the out-of-phase level is increased, the image will soon collapse and appear to come from inside the head.

With a really complex stereo signal, tweaking the side level can give dramatic dimensional effects. It is also a means of creating a fake stereo signal from a mono original.

M&S

An alternative way of dealing with stereo is as an M&S signal, the letters standing for Middle and Side. They can be converted back and forward. Technically

$$(L + R) \div 2 = M$$
$$(L - R) \div 2 = S$$

Equally,

$$(M + S) \div 2 = L$$
$$(M - S) \div 2 = R$$

What the '$\div 2$' means is a matter of controversy. The long-standing BBC convention is that it represents 3 dB (half power) of attenuation. An alternative view is that it should represent 6 dB (half voltage) of attenuation.

Faking stereo

You sometimes need to fake a stereo sound from a mono source. This can range from a unique background sound to a period train, only ever recorded in mono, passing across the sound stage between the speakers.

There are several techniques open to the balancer. With continuous sounds such as applause, trains, atmospheres, etc., it is often possible to get a spread effect by playing two copies of the same (or different) sources. One is panned left and the other panned right, with perhaps a third in the centre. If they get into close synchronization then compatibility problems may arise with phasing on a mono radio receiver. In *Cool Edit Pro*, this is easily done by using duplicate copies on different tracks and sliding the tracks relative to each other to get the right effect.

A better way, which gives no compatibility problems, is to play one source in mono and the other phase-reversed so that it only contributes to the stereo signal. This bonus sound is not heard by the mono listener, as it cancels out.

Again duplicate tracks are used, except that one (not both) of the duplicate tracks is converted to a stereo wave file and one channel (say the right) is phase-reversed (inverted). For a discrete sound, like an aircraft taking off or a train passing, sliding this track to delay the sound by 50–100 ms can produce good results, especially if the volume envelope on this difference channel is manipulated.

If you want a directional sound, like a train passing between the speakers, you need to do this in two stages. First generate a 'spread' train and mix this down to create a new stereo wave file. This can then be panned using the pan envelope control.

With atmospheres, very long delays can be used – try 30 seconds (if the FX is long enough). Changing the level of the out-of-phase signal in sympathy with the sound can improve realism.

Thunderclaps can be particularly effective if you run different mono claps on the in-phase and out-of-phase tracks. It can be difficult to find genuine stereo thunder that rolls around the room as effectively.

The same techniques can be used to beef up stereo sounds that do not come up to your expectations. Provided the sound is OK for any mono listener (if the recording is to be broadcast), then anything you add out of phase to your mix will be a bonus to the stereo sound without affecting the mono. If you can guarantee your recording will never be heard in mono, then compatibility is not a problem. Remember that in addition to mono radios, there are mono gramophones and mono cassette players.

12.9 Special transforms

The Effects transform special menu option (Figure 12.46) offers convolution and distortion options, which can be useful for producing science fiction distortions.

Weird effects for science fiction can be had aplenty, as virtually all the effects transforms, at extreme settings, produce very odd sounds – which hopefully only happen when you want them!

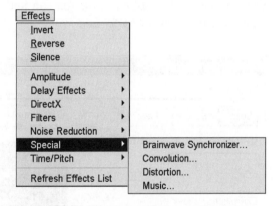

Figure 12.46 Effects transform special menus

Brainwave synchronizer

This is an idiosyncratic transform that makes claims about being able to process sound for listening on headphones. Syntrillium say that:

> The Brainwave Synchronizer can produce files that, when listened to with stereo headphones, will put the listener into any desired state of awareness. For example, by listening to 'waved' files, you can achieve states such as deep sleep, theta meditation, or alpha relaxation. Because of the nature of this function, it only works on Stereo waveform data, and to be effective, it must be listened to with stereo headphones. The Brainwave

function spatially locates the audio left and right, in a circular pattern over time. In order to spatially encode the signal, either the left or right channel is delayed so that the sounds will appear at each ear at different times, tricking the brain into thinking they are coming from either side. When this is done at frequencies of 3 Hz and above, the brain will start synchronizing at the same frequency, increasing its output of Delta, Theta, Alpha, or Beta frequencies.

Convolution

Convolution (Figure 12.47) is the effect of multiplying every sample in one wave or impulse by the samples that are contained within another waveform. In a sense, this feature uses one waveform to 'model' the sound of another waveform. The result can be that of filtering, echoing, phase shifting, or any combination of these effects. That is, any filtered version of a waveform can be echoed at any delay, any number of times. For example, 'convolving' someone saying 'Hey' with a drum track (short full spectrum sounds such as snares work best) will result in the drums saying 'Hey' each time they are hit. You can build impulses from scratch by specifying how to filter the audio and the delay at which it should be echoed, or by copying audio directly from a waveform.

To get a feel for Convolution, load up and play with some of the sample Impulse files (.IMP) that were installed with *Cool Edit Pro*. You can find them in the /IMPS directory inside the directory where you have installed *Cool Edit Pro*.

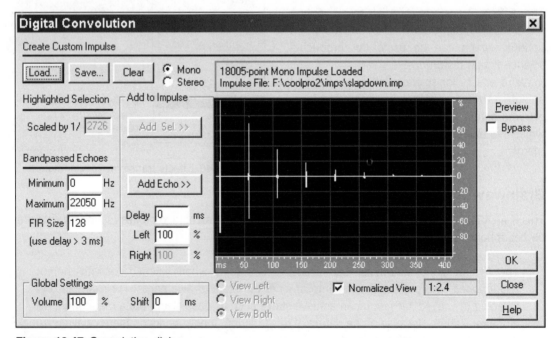

Figure 12.47 Convolution dialog

With the proper impulses, any reverberant space can be simulated. For example, if you have an impulse of your favourite cathedral and convolute it with any mono audio (left and right channels the same), then the result would sound as if that audio were played in that cathedral. You can generate an impulse like this by going to the cathedral in question, standing in the spot where you would like the audio to appear it is coming from, and generating a loud impulsive noise, like a 'snap' or loud 'click'. You can make a stereo recording of this 'click' from any location within the cathedral. If you used this recording as an impulse, then convolution with it will sound as if the listener were in the exact position of the recording equipment, and the audio being convoluted were at the location of the 'click'. Most of *Cool Edit*'s Reverb transforms are based on convolution algorithms.

Another interesting use for convolution is to generate an infinite sustained sound of anything. For example, one person singing 'aaaaaah' for 1 second could be turned into thousands of people singing 'aaaaaah' for any length of time by using some dynamically expanded white noise (which sounds a lot like radio static).

To send any portion of unprocessed 'dry' signal back out, simply add a full spectrum echo at 0 ms. The left and right volume percentages will be the resulting volume of the dry signal in the left and right channels.

Distortion

We spend most of our time trying to avoid distortion, but there times when we want it. This can be for dramatic purposes, or for things like a 'fuzz' guitar. This transform (Figure 12.48) does this by 'remapping' samples to different values.

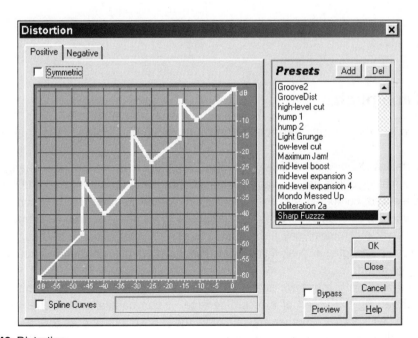

Figure 12.48 Distortion

Music

Music (Figure 12.49) is a simple sampler. You highlight a sound, and that is used to play the music that you enter into the dialog. Your highlighted range is treated as a crotchet (quarter note). If no range is selected, *Cool Edit Pro* will use the data on the clipboard. Note that the clipboard data will be filled with your sample automatically once music is generated; thus, selecting music a second time will automatically use your last sample. When you are ready, simply drag the notes and rests you desire to the music staff.

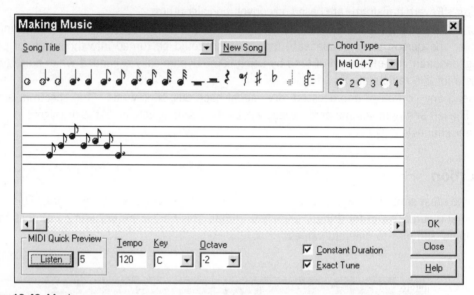

Figure 12.49 Music

12.10 Time/pitch

Cool Edit Pro has three effects transforms to do with pitch changing and time stretching (Figure 12.50).

The first creates Doppler shifts (Figure 12.50a). Doppler shifts are the pitch changes you hear when a noise-making object passes you. As it approaches the pitch is increased and once it passes the pitch is reduced, compared with its pitch when stationary. Perhaps the most common time we hear this is when an emergency services vehicle speeds by sounding its siren. This transform allows you to take a static recording of, say, a siren, and process it so that it sounds as though it is moving. The volume can be handled as well, if you want. The transform can also handle circular movement like a merry-go-round.

The second time/pitch transform is the pitch bender (Figure 12.50b), and this is the direct equivalent of varying the speed of a recording tape. It can handle a fixed change, or make a transition from one 'speed' to another. You can draw the curve of how it does this. As such, it forms the ideal basis to repair a recording made on an analogue machine with failing batteries. When the

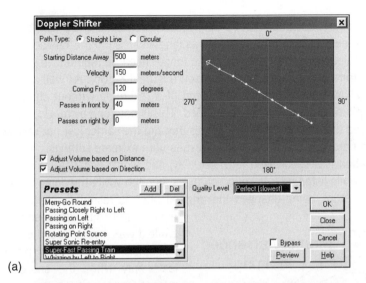

Figure 12.50 (a) Doppler shifts; (b) Pitch bender; (c) Stretch

tape speed varies on an analogue tape, as the speed decreases so the pitch decreases. If the tape is running at half speed, everything takes twice as long and the pitch is one octave lower. However, with computer manipulation this relationship no longer need be maintained.

Cool Edit Pro's third effects transform is Stretch (Figure 12.50c). As well as more conventional manipulations it allows you to change the speed of a recording without changing the pitch, or, alternatively, the pitch without changing the speed. This is not done by magic, but by manipulation of the audio. Indeed there were analogue devices that did the same, just not very well. Even in the digital domain you are less likely to get good results with extreme settings.

Pitch manipulation is a specialized form of delay, and works because, as in the chorus effect, changing a delay while listening to a sound changes its pitch while the delay is being changed. As soon as the delay stops being changed the original pitch is restored, whatever the final delay.

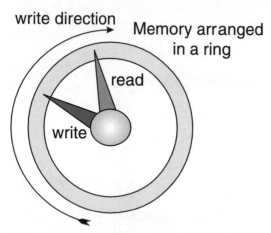

Anyone who has varispeeded a tape recorder while it was recording and listened to its output will know the temporary pitch change effect lasts only while the speed is changing. It returns to normal the moment a new speed is stabilized.

A digital delay circuit can be thought of as memory arranged in a ring, as in Figure 12.51. One pointer rotates round the ring writing the audio data, while a second pointer rotates round the same ring reading the data. The separation between the two represents the delay between the input and output. In a simple delay, this separation is adjusted to obtain the required delay.

Figure 12.51 Digital delay and pitch change

In a pitch changer, the pointers 'rotate' at different speeds. If the read pointer is faster than the write pointer, then the pitch is raised. If the read pointer travels slower than the write pointer, then the pitch is lowered.

Quite obviously there is a problem each time the pointers 'cross over', when a 'glitch' occurs. The skill of the software writer is to write processing software that disguises this. The software either creates or loses data to sustain the differently pitched audio output.

Uses

These transforms can be used for:

1 *Time stretching*. A radio commercial lasting 31 seconds can be reduced to 30 seconds by speeding it up by the required amount while maintaining the original pitch. Singers with insufficient breath control can be made to appear to be singing long, sustained notes by slowing down the recording and pitch changing to the original.
2 *Pitch changing*. The pitch change can be useful for correcting singers who are out of tune.

3 *Voice disguise*. These devices are sometimes used to disguise voices in radio and television news programmes. There was also a fashion for them to be used for alien voices in science fiction programmes.

4 *Music*. With judicious use of feedback within the unit, pitch changing can be used to thicken and enrich sounds.

Basic effects

There are three basic effects at the top of the Effect menu drop down: Silence, Invert and Reverse.. Although they are very simple, they are extremely useful.

Silence

This mutes the selected section of the wave file. Its main purpose is for music balancing, as you can use it to remove spill, coughs and sneezes from, say, a vocal track. This means that you do not have to set the volume envelope for each vocal entry and exit, but can guarantee absolute digital silence in breaks in the performance.

Don't confuse this with Generate/silence. The difference between the two is that this transform mutes a section of existing audio, making no change to its length. Generate/silence inserts a given number of seconds of silence at the cursor, so that inserting 10 seconds of silence will make the file 10 seconds longer.

Invert

The Invert menu option is a fairly specialized effects transform. All it does is invert the waveform so that positive becomes negative and negative becomes positive.

In music balancing, this is the equivalent of the phase reverse switch on a mixing console. You can find, when you come to a mix down, that a microphone was out of phase with others used at the same time. Microphones should be in phase, which means that their output voltages are going in the same direction; an impulse that makes the diaphragm move inwards on the microphones produces the same direction voltage. All the microphones should produce a positive voltage, or all the microphones should produce a negative voltage – it doesn't matter which, as long as they are the same. If they are out of phase they tend to cancel, and the mix can sound a bit like a drainpipe.

The left and right channels of stereo should be in phase too, as otherwise a central image seems to be coming from inside the head rather than from between the speakers. Worse, if the recording is broadcast, people listening on mono receivers will not hear anything in the centre as it will cancel out. See the description of *Cool Edit*'s phase analysis (page 187) for more information.

Reverse

Reverse does just literally that; it reverses the sound so that the beginning becomes the end. Its main uses are for producing reversed echo and alien voices. It can also be used in batch files to reduce phase shift errors using filters by running the filter, reversing the audio, running the filter again so that any phase timing errors are reversed, and then reversing the audio back to normal again.

Reversed echo is often used for magical effects: witches and aliens. It sometimes has a role in music, although it is an entirely unnatural effect.

A multistage process is used. First the programme material is reversed, and then it has a reverberation effects transform applied to it. That transformed recording is then reversed again, which restores the original speech etc. to the right way round, but with the reverb now reversed – what was a reverb die-away now becomes a build-up. Satisfyingly weird effects can be created this way. However, the relative balance is critical, so again it is best if the reversed reverb is created as a 'wet-only' track so that it can be mixed with the original while you can actually hear the intended effect.

If you have a performer who is a good mimic, you can create a really good alien voice by getting the performer to record the lines and then reversing them. The performer now imitates the sound of the reversed speech – this takes some practice – and is then recorded imitating the backwards speech, and then that recording itself is reversed. All being well, an inhuman voice will be heard speaking very strangely but intelligibly. This sort of trick has to be done in relatively short takes to work well.

12.11 Multitrack effects transforms

The multitrack view of *Cool Edit Pro* has three effects transforms of its own. They are here because they generate a new track or tracks in the multitrack view.

The envelope follower and the vocoder combine, non-destructively, two different audio tracks to create a third. It can be very frustrating to try to use these, as the menu options are greyed out unless the conditions are right. To use *Cool Edit Pro*'s vocoder or envelope follower, you have to both select two tracks *and* highlight the section to be processed. The easiest way to do this is to have the control and process tracks adjacent to each other, and to highlight both with one action. Place the mouse cursor over the first track, press and hold down the left mouse button at the beginning of (or just before) the wanted section, and drag the cursor down to the next track and along to the end of the section required. You can do this starting from the end and moving to the beginning if you prefer. This is often the best way if you are starting at the very beginning (0:00.00). Both tracks will now be highlighted, and the Effects/vocoder and Effects/envelope follower menu options will be enabled (this happens only if two tracks are selected).

If the two tracks are not adjacent, you can still do the same, except that the tracks in between will also be enabled. You need to deselect these individually by holding the Control key down and single-clicking on each.

Selecting the section and then enabling two tracks by holding the control key down and single-clicking on them will also work.

Envelope follower

The Envelope follower (Figure 12.52a) varies the output level of one waveform, based on the input level of the other. The amplitude map (or envelope) of one waveform (the analysis wave) is applied to the material of a second waveform (the process wave), which results in the second waveform

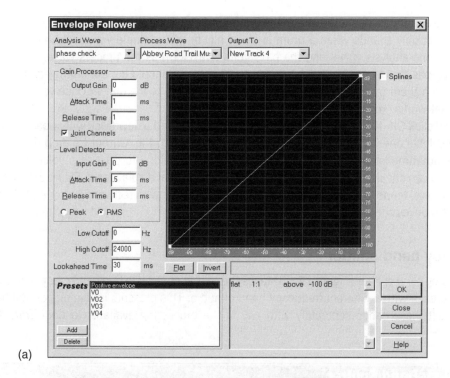

Figure 12.52 (a) Envelope follower; (b) Inverting the line in the graphical display

taking on the amplitude characteristics of the first waveform. This lets you, for example, have a bass guitar line that only sounds when a drum is being hit. In this example you would have the drum waveform as the analysis wave, and the bass guitar waveform as the process wave.

The dialog has a go at guessing which way round the tracks should be, but often gets this wrong. You can switch them by clicking on the drop-downs at the top of the dialog. Once you have found a setting that works, you can save it as a preset.

When you click OK, the normal progress dialog will appear and a new track will appear. Normally the next available one will be used, but you can change it with the 'Output to' drop down.

There is no immediately obvious way of inverting the envelope to provide a voice-over effect. However, you can cheat this by inverting the line in the graphical display (as in Figure 12.52b). The dip in the 'music' tends to anticipate the voice and be a little slow in recovering after it. A 3-second look ahead reduces the gap.

Frequency band splitter

The frequency band splitter (Figure 12.53) takes the output of a single track and creates up to eight new tracks, each with a different frequency band of audio. This presents intriguing possibilities, as you can treat these bands separately with any of the processing available to *Cool Edit*. At the

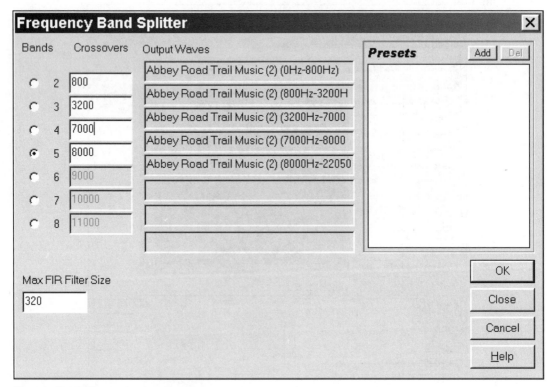

Figure 12.53 Frequency band splitter

simplest, you could boost the level of a cymbal to change the sound or you could even compress it. By putting compression on each track you have the basis of a multiband compressor for mastering. This is one of those tools that, while appearing simple, opens up a host of possibilities.

You can have up to eight bands, and you can change the crossover frequencies by typing them in the left hand boxes. The number of tracks produced is selected by the buttons at the side; these are not toggles, but clicking one will produce that number of bands.

Vocoder

A vocoder (Figure 12.54) is a special effect that can make inanimate objects appear to speak or sing. If you want the leaves of a tree to talk or sing, then the vocoder may do what you want. It takes one sound (the process signal) and modulates it with the second (the control signal). Modulate a telephone bell with someone saying 'ring ring', and you could have a talking telephone bell.

The top three boxes tell you about the two input tracks and where the transform will be placed. The left and centre drop-downs allow you to set which track is the control wave and which is the process file. The third box displays the next track empty at the selected area where the resulting transform will be placed.

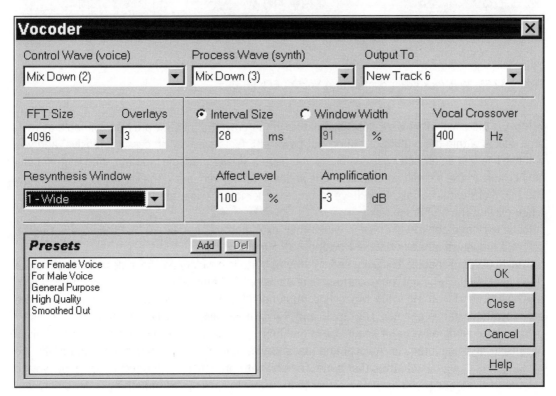

Figure 12.54 Vocoder

Some presets are provided, and the help file advises:

> For ease of use, try checking 'Window Width' and set it to about 90 per cent, use three or four overlays, a resynthesis window of 1 or 2, and FFT sizes from 2048 to 6400.

A fair amount of fiddling must be expected to get good results.

Real-time controls and effects

In the multitrack editor, you have various controls in the left-hand column. Which ones you see are controlled by the tabs at the top labelled 'Vol', 'EQ' and 'Bus' (Figure 12.55).

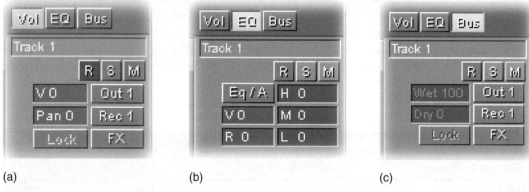

(a)   (b)   (c)

Figure 12.55 Multitrack editor controls: (a) Vol; (b) EQ; (c) Bus

Most of the effects that were described earlier in the Wave Editor view are also available as real-time effects within the multitrack editor. But beware! Effects are very processor intensive, and you will need as fast a computer as possible. If you have a slower machine, you may find that you will only be able to use a small number at a time and will, having rehearsed the setting to get them how you like them within a mix, then have to process the actual wave file to reduce processing load when playing the multitrack mix.

If the left-hand column is dragged wider, then extra controls appear so that they overlap in the different tab views albeit in different positions (Figure 12.56).

The 'R' (Record enable), 'S' (Solo) and 'M' (Mute) buttons have already been described in Chapter 9, as have the volume and pan controls, here showing 'V0' and 'R0'.

The 'Lo', 'Mid' and 'Hi' value windows control individual track parametric equalizers, which are described later in this chapter. The 'Lock' and 'FX' buttons give you access to real-time effects.

In an ideal world all your FX should be in real time, just like on an ordinary sound mixer. This is because everything interacts. A vocal line that sounded fine may need hardening if you make the backing music brighter. On the other hand, the ability to pre-record your effects does allow you to use the slow computer you have now rather than having to save up for something faster before you can do anything.

(a)

(b)   (c)

Figure 12.56 (a)–(c) Expanded track controls

A useful compromise, especially for reverb, is not to alter the original wave file but to generate a new wave file that is just the reverb for a track. Controlling the level of this additional track requires much less processing power.

Cool Edit Pro has a neat half-way house. You can 'lock' the effects once you have a setting that you like. There will then be a pause while the progress bar shows it processing and saving the FX. It can be left untouched and the processing overhead will be small, but a click of a button will restore the track to real-time effects.

The FX transforms used by the wave editor are contained within a 'wrapper' of extra controls needed by the multitrack view, while some of the action buttons (like preview) are hidden as they are not appropriate for the real-time nature of the effect. The appearance is often slightly different, with things like preset windows moved to fit more easily into the wrapper. The flat controls of the wave editor are replaced by 3D ones in the real-time version (Figure 12.57).

The wrapper places each effect as a tabbed window within its dialog (Figure 12.58). The 'serial' and 'parallel' buttons control whether the effects feed in to each other (left to right) or separately contribute to the sound. For example, you may use the delay effects serially to feed the reverb, to get extra separation of the effect from a vocal driving it. Equally you might be applying both flutter echo and reverb, and want them in parallel so that they appear separately on the mix.

'Rack setup' (Figure 12.59) is used to modify the processing on a track. This same dialog is called the first time you click on the Track effects by clicking on the track's FX button. You can alter the order of the effects, which can be important if you want to feed them serially.

Rack setup, FX setup and FX mixer are conveniently called up by right-clicking on the FX button. It also has a Bypass option, so all the effects on that track can be taken out to that you can compare with the original. Individual effects can be bypassed on the FX mixer.

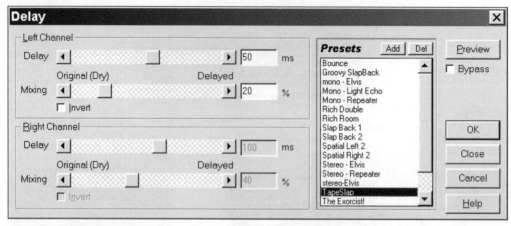

(a)

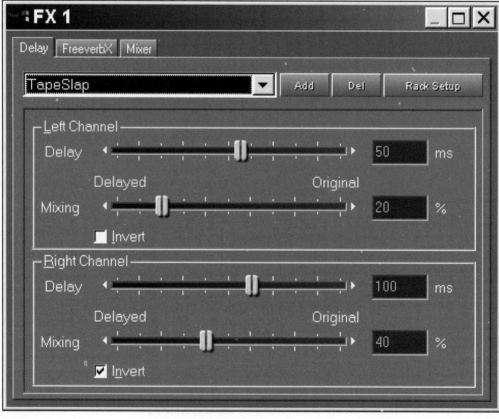

(b)

Figure 12.57 Delay effects transform (a) as it appears in the wave editor, and (b) as it appears in the multitrack view. It has the buttons on the right of the waveform view version. The Presets window has been rearranged. It appears as a tabbed page of the FX window of the track.

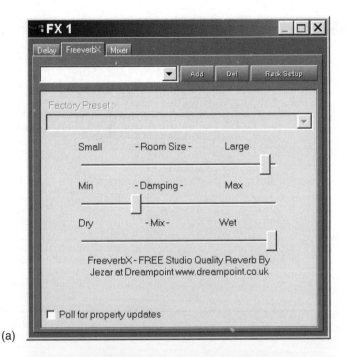

Figure 12.58 (a) FreeVerb Third Party Effects transform as it appears in a tabbed page of the FX window of the track. (b) The mixer window allows the setting of the amount of direct (unprocessed or 'Dry') and indirect (processed, 'Wet') sound. The serial and parallel buttons control the routing of the Effects, so that one can feed the other or they can contribute independently

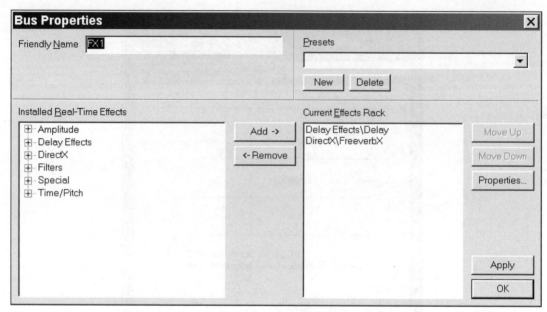

(a)

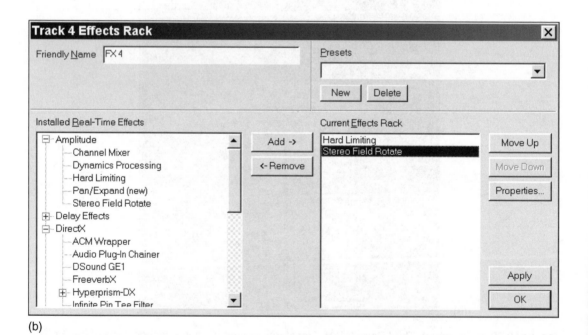

(b)

Figure 12.59 Bus properties dialog, also know as 'Rack setup' or Track effects rack, showing (a) Effects submenus unexpanded; (b) Effects submenus expanded

Envelope

Some editors provide you with mix automation by moving the controls on a graphic representation of a mixer. While it is great fun watching faders move, apparently of their own accord, it tells you little about the overall mix, and you suffer from the 'one-finger' problem of having to control everything with the mouse.

Cool Edit Pro uses an alternative way of automating the mix; using envelopes. We have already met the basic ones, which are pan and track level, in Chapter 9. For a full-scale mix there are more; wet and dry and FX parameter envelopes.

The Wet/Dry parameter envelope allows you to have not only different amounts of effects on different audio blocks in a track using real time effects but also allow you to change them from moment to moment. You have to ensure that View/Enable envelope editing is ticked, and that you have selected View/Show Wet/Dry mix envelopes. You may well want to turn off the display of other envelopes to avoid cluttering the track display. When the envelope is at the top of the block you get 100 per cent 'wet' – that is, all effects and no direct sound. At the bottom you will get all direct sound but no effects. If View/status bar/Data under cursor is ticked, then as you adjust the envelope you can see a numerical value for what you have set in percentage points.

Do not confuse the Wet/Dry envelopes with the Wet/Dry windows at the left of the track. These are used with output buses described later in this chapter.

The FX parameter envelopes appear when you apply some *Cool Edit Pro* effects to a track. These include Pan/Expand, Dynamic EQ and Dynamic delay. When added as real-time effects, their dialogs will have an 'automated' check box in their settings dialogs. There can be more than one parameter controlled with some effects. For example, Dynamic delay allows you to vary delay and feedback over time. Dynamic EQ changes it settings dialog appearance totally (Figure 12.60a) and allows you to change gain, frequency and 'Q' entirely from the parameter envelopes. Figure 12.60b demonstrates the 'Data under cursor' window showing the 'Q' value.

The Tracks mixer window

Version 2 of *Cool Edit* has a large graphical mixer available, and this can be used to set the basic track levels. It corresponds to the controls at the left hand side of each track. Each row of controls can be removed if they are not need to be seen. The mixer reformats itself depending on how large its window is (Figure 12.61).

This is fine if you have a sentimental attachment to the idea of mixing desks, but actually provides you with much less information that the envelope controls do. The envelopes can show you at a glance exactly what you have done to a mix. Using a mouse, you only have one finger and setting mix can be slow. Irritatingly, the mixer is the only place, on screen, that you can set the master gain. You can resize the mixer window to be tall and thin to show only the master gain control and dock the window to the left of the tracks without losing too much screen space.

The Bus mixer window

The Bus mixer window (Figure 12.62) allows you to set up 'groups', as they would be known on a mixer. Instead of the output of a track being selected to a particular sound card output it is, select

(a)

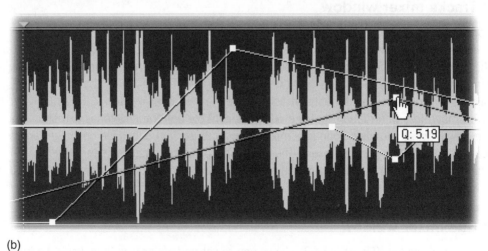

(b)

Figure 12.60 FX parameter envelopes: (a) Dynamic EQ; (b) Data under cursor

(a)

(b)

Figure 12.61 *Cool Edit Pro's* Mixer Window (a) in a large window; (b) in a small window

Figure 12.62 Cool Edit Pro's Bus Mixer Window in a large window

to a bus fader. Just as in its public transport use, 'bus' is a contraction of 'omnibus'. The original term 'omnibus bar' was first coined in power stations for the circuit that combined the outputs of several generators. Because of the currents involved this was literally a bar. This was later taken up by the audio industry, although the 'bar' was now a thin strip of copper on a circuit board. Just as 'omnibus transport' meant 'transport for all', so the bus bar is a circuit for all. It applies generally to a part of a circuit that has several inputs and/or outputs.

Although the Bus mixer window will always show at least a single channel, you have to create a bus by clicking on the 'New' button. Effectively the rightmost channel is a dummy used to create new buses. Clicking on the 'Out' button at the left of the track gives you the option of selecting directly to a sound card output or to a bus. Figure 12.63 shows three buses have been created. The 'Out' buttons on the bus fader normally feed to the first sound card on *Cool Edit*'s list, but clicking them allows you to change this. You cannot, at the time of writing, feed the output of a bus to another bus, but only to a sound card.

OK, why would you want to do this? The answer is sub-mixing and processing. There are some sources in a music balance that inherently have to be internally balanced. The classic example is a drum kit, which is likely to have several mics to balance it. Once you have achieved a good internal balance, it is very convenient to be able to move that mixed entity up and down in the main mix. On a mixing desk it would be very difficult to grab all the drum faders and move them up and down together. So it is on *Cool Edit*; you only have the one 'finger' of the mouse. Instead you can set up a drum bus and use the bus fader as a drum fader. Unfortunately you do not have any envelope automation on the bus level, but this can still be a help.

Processing is the second reason for using a bus. It is quite usual for several tracks to need the same processing, such as reverberation. If each track opens its own reverberation trans-form on its 'rack', then that will make a big dent in the processing requirements for real-time

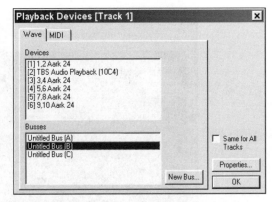

Figure 12.63 Three buses have been created

playback. Instead put that reverb into the bus mixer, and then only one tranche of processing is required. While this seems an altogether good idea, there is still a niggle. Each track may well want the same acoustic and therefore the same reverb program, but it is very likely that different wet/dry mixtures will be needed. This is what the wet/dry windows in the track control box at the left of the track are for – they allow you separately to control the wet/dry mixture for each track while using a common processing transform.

Track equalizer

Each track comes with its own real-time equalizer. At first glance it looks very simple, with three windows to control the low, middle and high frequencies. Like the other variable controls, these are changed by pressing down the left mouse button and dragging the mouse up or down. However, if you double-click on any of these windows a fully featured parametric equalizer appears, complete with graphic display. In fact there are two alternative equalizers, 'A' and 'B', which you can switch between easily by single-clicking the 'Eq' button. It will toggle between showing 'Eq/A' or 'Eq/B'. At its simplest, this allows you to use 'Eq/B' as a flat reference to switch the equalization in and out. Of course, you can also use it to switch between alternative equalizations. Sometimes the alternative equalization will be an 'equalized flat'. For example, you may already have established that you need a high-pass filter setting on the low control, but now want to try a bit of middle boost on top of that. You can copy a setting from one EQ to another by double-clicking on the 'Eq' button. If it is showing 'Eq/A' then the 'A' equalizer setting will be copied to the 'B' equalizer, and the reverse happens if you click on 'Eq/B'.

The low and high controls can be switched between shelf and band equalization (Figures 12.64, 12.65).

Shelf equalization operates like an ordinary bass or treble control on a hi-fi. However, unlike an ordinary hi-fi, you can alter where the 'turnover' from flat to sloping takes place. The boost out cuts levels out after a couple of octaves so as not to excessively boost extreme high or low frequencies.

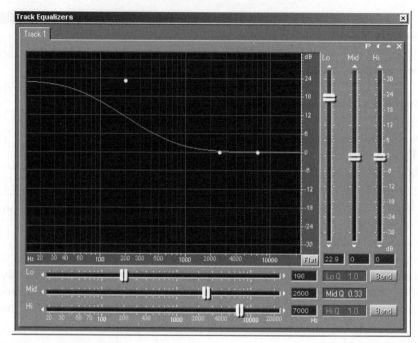

Figure 12.64 Low frequency shelf setting

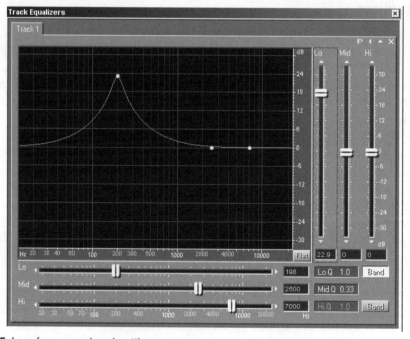

Figure 12.65 Low frequency band setting

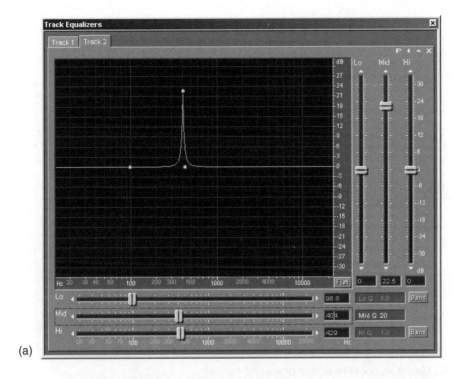

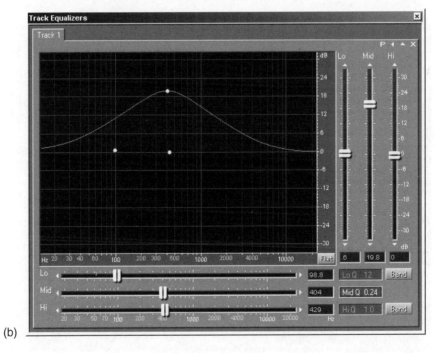

Figure 12.66 (a) Middle control high Q setting; (b) Middle control low Q setting

PC Audio Editing

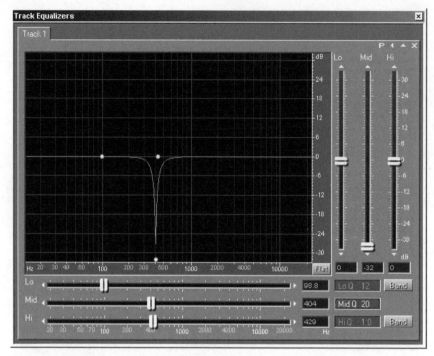

Figure 12.67 Maximum boost to isolate unwanted frequency

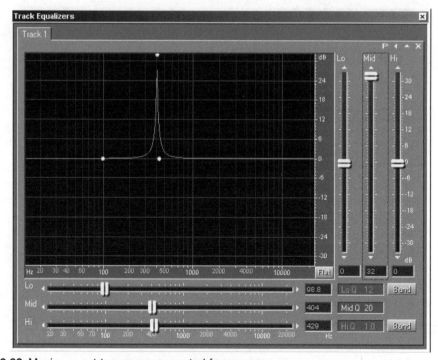

Figure 12.68 Maximum cut to remove unwanted frequency

Clicking the 'Band' button changes the filter into a band-type equalizer that operates like a middle control. Thus you can have three middle controls, which can be much more useful than shelf equalization. The middle is, as you would expect, always set to Band. Any of the bands can be adjusted for Q. Any of the controls can always be controlled for the active frequency they start operating.

Semi-professional mixers sometimes have just a simple plus or minus control for the middle. Better ones have a sweep control where you can adjust the frequency of boost or cut. The very best also have a 'Q' control that adjust the width of the boost or cut (Figure 12.66). The very narrow setting can be useful for removing interference sounds that have been picked up ranging from buzzes to lighting whistles (but don't forget the notch filter is already set up to reduce mains electricity related buzzes).

A good technique to remove a single frequency is to set the band control to maximum Q, the narrowest setting. At maximum boost, move the frequency control until the unwanted sound is found and has become much louder (Figure 12.67). Now move the level control from maximum boost to maximum cut, and the unwanted frequency should be tamed or eliminated (Figure 12.68). In practice, you may find that a maximum setting will affect the overall wanted sound as well as removing the unwanted. Here, a compromise has to be made. Only your ears can tell you how much.

13

Reviewing material

13.1 Check there are no missed edits

In news and current affairs broadcasting, many compromises have to be made. Elsewhere, there is less excuse. Traditionally, with quarter-inch tape, you would spool back from the end, with the tape against the heads, controlling the speed so that you could hear the speech rhythms. A missed edit, a gap or talkback, would show as a break in the rhythm. Any missed edits that were found could be razor-bladed in seconds.

With material prepared on an audio editor, life is not so simple. If the final item is transferred to CDR or DVD, there is no way of listening to the recording while this is done. Transferring to DAT involves a real-time copy, which can provide a good opportunity to review the item.

The snag is that if a missed edit is found then the copy has to be restarted from the beginning. Ironically, this can lead to a programme being prepared entirely using digital technology, only to be transferred to analogue quarter-inch tape so that any missed edits can be edited the old-fashioned way rather than having to start the transfer from the beginning.

13.2 Assessing levels

Very rapidly, the eye becomes used to assessing levels from the appearance of the waveform display. However, there is no substitute for listening to the item in one go, on good equipment.

You should also make a habit of listening to your material in the environment that it will be used. If you are making a promotional cassette to be heard in a car, then listen to your tape in a car – preferably a noisy one – so that you can check that levels don't drop to inaudibility.

If the item is to be used on the telephone, as an information line recording, then this presents a virtually unique balancing challenge. This is genuine monaural sound: one-eared. Check your balance with headphones worn so that only one ear is used. If you have facilities to feed the audio down a telephone line, then do so. Remember, even if you have been asked to provide the material as analogue audio, it will almost certainly be reduced to, at best, 8-bit digital audio on the play-out device. Lower bit rates are often used, with the resulting quality being much below what the telephone system (which is an 8-bit system) is capable of.

13.3 Listening on full quality speakers

This will reveal every deficiency in your recording. It becomes a production decision as to what the listener will hear the recording on. It is as important to listen on poor speakers. Music studios always have a set of small speakers to check balances on low bandwidth speakers. A good balance travels well. If your item sounds fine on good speakers, and on a portable cassette player, then it is likely to be successful.

The 'grot' speaker also reveals whether effects that sound impressive on big speakers or on headphones also work on average equipment. A classic mistake is to use a deep bass synthesized sound for a dramatic heartbeat. While sounding effective on big speakers with extended bass response, the same sound may appear as low level clicks on a transistor radio.

There is a definite difference between 'hi-fi' loudspeakers and 'monitoring' speakers (the speakers that come with PCs are neither). Hi-fi speakers are designed to make everything sound good and well balanced, whereas monitoring speakers are designed to be analytical. They are your equivalent of the doctor's stethoscope. You want to be able to hear things, and correct them, before your listeners hear them. This is why sound balancers monitor at higher levels than ordinary mortals – they wish to hear inside the balance. They will also 'dim' the speakers' (reduce the level by 12 dB to 15 dB) from time to time to check realistic levels on both their monitoring speakers and the 'grot' speakers.

14

Mastering

14.1 Line-up and transmission formats

The mastering process is concerned with producing the final product that is actually going to be used. This could be for use by yourself – say, a CD of audio illustrations for a talk – or it could be a multitrack tape for a *Son et Lumière*. It could be a programme for broadcast. Here, the final recording has to be prepared in the correct way to fit in with the requirements of the broadcaster as to levels and line-up information.

The requirement will differ between organizations and reflect their internal practice. For example, if your customer requires programmes provided on compact disc to be recorded with a peak level of −4 dB, then all you need to match this requirement is to normalize your programme to −4 dB and burn it to CD.

Increasingly broadcasters are accepting material on CD-ROM as a .WAV file. They usually require that the professional sampling rate of 48 kHz is used for these. Wave files can also contain formatted text information. *Cool Edit Pro* provides dialogs for entering this (Figure 14.1).

There is also a European Broadcasting Union standard for including text in the wave file to provide production information (Figure 14.2).

Line-up

Quarter-inch tape and DAT usually need to have a line-up tone, or tones, at the front of the programme. Here, BBC practice is for 20–30 seconds of, nominally, 1-kHz tone on both tracks recorded at 11 dB below peak programme level. Analogue tapes are also required to have a section of 10-kHz tone, so that head alignment errors can quickly be detected.

The 'Generate' menu title covers useful tools here.

14.2 Generate silence

This option does just that; it generates a set number of seconds of silence. It is different from the silence in the Effects section, which mutes the highlighted section of the recording. Here, the silence is generated as a new recording – any audio recording after the cursor is pushed onwards to make way for the silence, rather than being replaced by the silence. This means that you can insert a silence between a line-up section and the programme material.

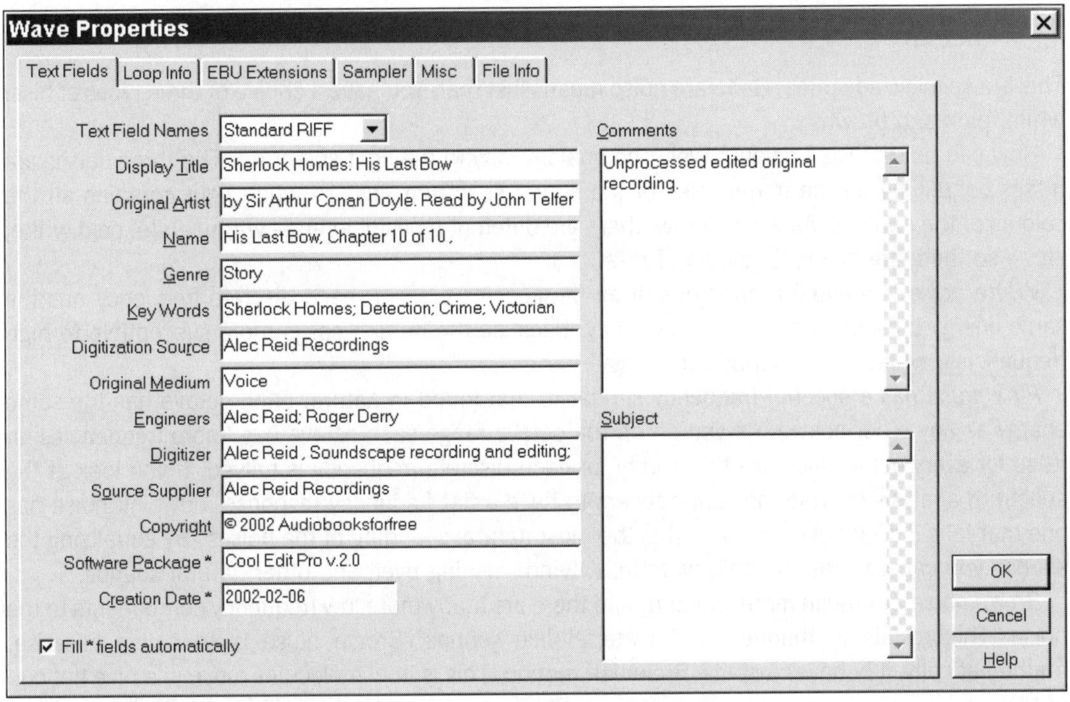

Wave Properties ☒

Text Fields | Loop Info | EBU Extensions | Sampler | Misc | File Info

Text Field Names [Standard RIFF ▼]

Display Title [Sherlock Homes: His Last Bow]

Original Artist [by Sir Arthur Conan Doyle. Read by John Telfer]

Name [His Last Bow, Chapter 10 of 10 ,]

Genre [Story]

Key Words [Sherlock Holmes; Detection; Crime; Victorian]

Digitization Source [Alec Reid Recordings]

Original Medium [Voice]

Engineers [Alec Reid; Roger Derry]

Digitizer [Alec Reid , Soundscape recording and editing;]

Source Supplier [Alec Reid Recordings]

Copyright [© 2002 Audiobooksforfree]

Software Package * [Cool Edit Pro v.2.0]

Creation Date * [2002-02-06]

☑ Fill * fields automatically

Comments
[Unprocessed edited original
recording.]

Subject
[ ]

[OK]
[Cancel]
[Help]

Figure 14.1 Wave text information

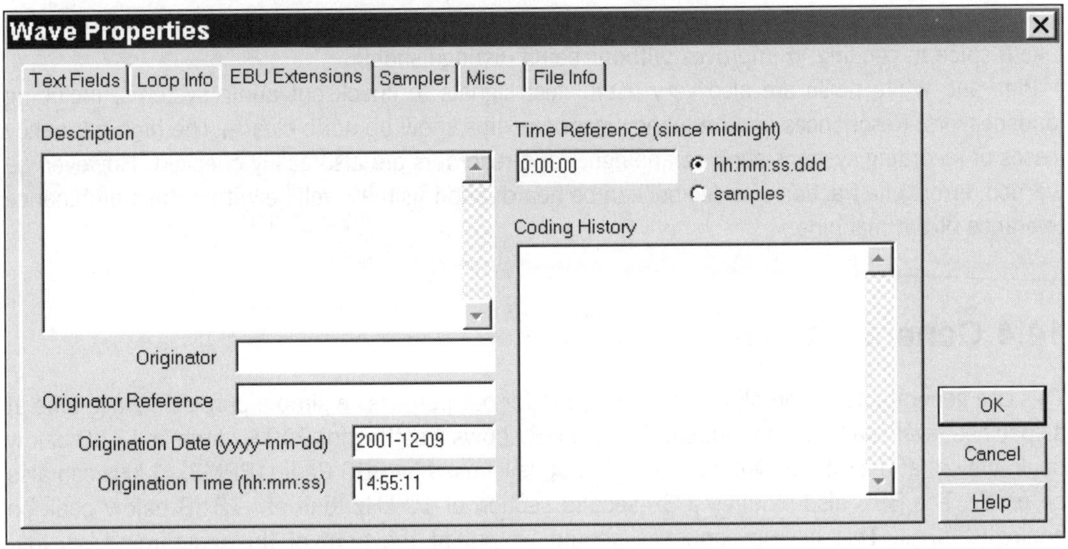

Wave Properties ☒

Text Fields | Loop Info | EBU Extensions | Sampler | Misc | File Info

Description
[ ]

Originator [ ]

Originator Reference [ ]

Origination Date (yyyy-mm-dd) [2001-12-09]

Origination Time (hh:mm:ss) [14:55:11]

Time Reference (since midnight)
[0:00:00]   ⊙ hh:mm:ss.ddd
            ○ samples

Coding History
[ ]

[OK]
[Cancel]
[Help]

Figure 14.2 EBU extensions

14.3 Generate noise

This is a specialized option. Generate noise means just that. You have a choice of three types of hiss: white, pink and brown.

How can noise have a colour? Well, the analogy is with the light spectrum. The three noises are hisses containing all the frequencies of the sound spectrum, just as white light contains all the colours of the rainbow. As we all know, there are different types of white – warm white, cold white, etc. – so there are different colours of hiss.

White noise has equal proportions of all frequencies present; each discrete frequency has the same energy present as any other frequency. Because the human ear is more susceptible to high frequencies, white noise sounds very 'hissy'.

Pink noise has a spectral frequency similar to that found in nature. Each octave has the same energy as any other octave. As you go up the octave range each octave has 'more frequencies' in it; so for every octave increase the energy for each discrete frequency is halved. If you look at the spectrum analysis of white noise it appears to have a flat frequency response, but pink noise has one that falls 3 dB for every octave. It is the most natural sounding of the noises. By equalizing the sounds you can generate rainfall, waterfalls, wind, rushing river, and other natural sounds.

Brown noise has much more low-end, and there are many more low frequency components to the noise. This results in thunder- and waterfall-like sounds. Brown noise is so called because, mathematically, it behaves just like Brownian motion. This is how molecules move in a nice hot cup of tea!

As has been suggested, these three types of hiss can be dragooned into helping a sound effects mix. With a little top cut, brown noise can take on the role of 'city skyline'. This was the background rumble that, for decades, was the sound of silence on television. When they had mute film, a touch of skyline gave a sense of something going on. Legend had it that it was the sound of Victoria Falls played at half speed.

This sort of rumble added to a sound effects mix can give it a depth it might not otherwise have. Like a spice in cooking, it improves without being distinguishable.

Pink and white noise are also very useful test signals to check out audio systems, including loudspeakers. Resonances and frequency response dips show up quite clearly. The high frequency losses of recording systems such as analogue tape recorders are also easily checked. However, be warned: errors of a fraction of a decibel can be heard, and this may well be within the maintenance tolerance of the machine.

14.4 Generate tones

This can generate quite complex tonal mixes, but for our purposes a simple pure sine wave tone at a specified level is all that is required. Figure 14.3 shows a setting for 900 Hz tone at 11 dB below peak, which, at the time of writing, is the line-up level used by BBC Radio (PPM3$\frac{1}{4}$, 1 kHz can also be used). The BBC also requires a 30-second section of 10-kHz tone at –18 dB below peak on analogue tapes. This line-up sequence should be put at the front of the programme on your computer so that if you are dubbing your final programme to quarter-inch, the whole sequence from

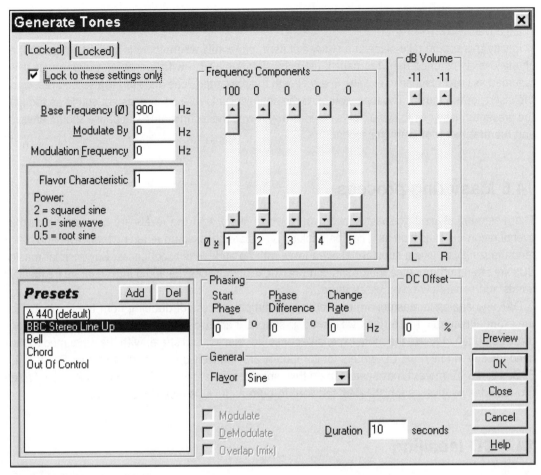

Figure 14.3 Generate tones

line up to end of programme is dubbed in one pass. For line-up the phase difference should be zero, as tone may be used as a system phase check.

The Generate tones dialog has the option of using beginning and end settings by unticking the 'Lock to these settings only' box, when the tabs showing 'Locked' become labelled Initial settings and Final Settings. This could be used to generate a continuous frequency response test run for 'squeaking' the programme chain.

14.5 Changeovers

If quarter-inch tape is being used, then the maximum time available on a $10\frac{1}{2}$-inch NAB reel is 30 minutes. Programmes longer than this have to be on more than one reel, and each reel has to be cut for a changeover in such a way that it can be performed perfectly without rehearsal.

While this is a rapidly disappearing requirement, it is still worth considering how best to cut a changeover.

Ideally this should take place at a change of item, preferably where there is a slight pause. (If the changeover is botched then the pause just gets slightly longer – with presented programmes it is regarded as good practice to end the first reel with the presenter's cue and start the second reel with the insert, or the music. The logic is that, if the changeover is missed, it does not sound as though the presenter has gone to sleep but is obvious that it was the technician!) The pause should always end the first reel, *not* start the second.

14.6 Mastering process

The mastering stage is the last chance you have to check that the levels are consistent and the overall recording is as loud as it can be. Picking off isolated peaks and reducing their level and then renormalizing can give a substantial extra level with no audible consequences. Commercial music CDs are often put through a multiband compressor, which divides the audio into separate frequency bands and separately compresses them.

DAT and analogue masters are created by the simple process of dubbing from the computer to the external recorder. Mastering CDs can be as simple, if an external CD recorder is used. However, these can often only use the more expensive consumer blanks, and share, with the other media, the need to copy in real time. CD writers for computers are relatively cheap, and can burn CDs at 2, 4, 6, 8, 12, 16 (or more) times speed. To do this, special software is required. Version 2 of *Cool Edit Pro* includes relatively sophisticated software to burn audio CDs (see Chapter 15).

14.7 CD labelling

Surprisingly, the most fragile surface of a CD is the label side, not the playing side. Scratches on either side can cause errors. Many CDRs come printed with labels for handwriting their contents. This is perfectly possible, but care must be taken to use a pen certified as suitable. A ballpoint pen will just damage the CD, and some ink solvents will dissolve the protective layer. Some CD blanks have a thin layer of ceramic on their label side that not only protects the disc but also makes them suitable to be directly printed using a modified inkjet printer.

Using ordinary stick-on labels may or may not work, depending on the glue used on their backing. However, they can end up damaging your drive rather than the CD, as the CD may be being spun at 40 or 50 times the standard speed and the off-centre weight can disrupt reading the disc and strain the bearings.

Purpose-made labels for CDs can be purchased. Different brands are usually associated with different proprietary applicators that will centre them properly. While these seem expensive, the applicators also come with software for designing and printing onto the labels.

You can also buy transparent labels to use with unprinted disc blanks to make the CDR label look more like a conventional CD. However, getting these onto your CD without air bubbles is more of an art than a science. Some people, who wish doubly to protect their CDs, put the transparent labels on top of the paper printed ones and also make them impervious to liquid spills, which make the inkjet printing run. Ensure that the labels you use are suitable for your printer, as inkjet and laser printers have different requirements. Modified inkjet printers that can print directly on to CD surfaces are also available.

14.8 Email and the World Wide Web

The requirements for the Internet are constantly changing, but the constant for all but those with expensive lease lines is that there is a limit to the amount of data that can be sent every second. What is now often called POTS, Plain Old Telephone System, can at the very best with a V92 modem get 6–8 kbytes per second of data, and even that is in one direction only – from the Internet Service Provider to the user. Upload speeds in the opposite direction are usually half that.

ISDN can offer those sorts of speeds reliably in both directions, or twice that if both channels of an ISDN circuit are combined (usually at the penalty of being charged as two calls). While weary broadcasters are apt to translate ISDN as the 'It Sometimes Doesn't Network', it actually stands for 'Integrated Services Digital Network'. It is capable of handling any data, whether telephone, fax, graphics or even stereo music. The main complications are due to a plethora of standards, many of which are not quite compatible with each other. Lease lines offer virtually any speed that you want, but at a price. ADSL (Asymmetric Digital Subscriber Line), sometimes known as 'Cable modem', allows you to be on-line all the time and have access to much faster data communications. Different cable and telephone companies differ in the 'contention ratios' they offer. The bandwidth you are offered is shared, and the amount you actually get from moment to moment is dependent on how many people are using the network.

However well you are equipped, you want audio that you put onto the Internet to be accessible to users who are not so well equipped. This means that there is a need to make files as small as possible with a minimum loss of quality. A CD quality recording takes up 10 Mbytes for every minute. This makes it impossible to play in real time from the Internet, as well as occupying a great deal of space on the hard drive of the server where it will be stored.

There are simple ways of reducing this, but these have quality penalties. Mono rather than stereo halves the size instantly, as does halving the sampling rate. Reducing the bit rate from 16 to 8 halves the size yet again. By the time you have done all this, the quality is getting decidedly iffy.

What is needed is some way of compressing the audio data in order to occupy less space. Immediately, ZIP files come to mind. They are almost universally used as a way of transferring other files. It makes the files smaller and they take less time to be sent. A ZIP file uncompresses its contents to exact copies of the originals. This works very well for text, but very badly for wave files. It can even happen that the zipped version of a wave file is actually bigger than the original.

What is needed is a different way of recording the data, to sound as good as possible, while throwing away as much data as is feasible. These are known as 'lossy' forms of compression, unlike the lossless ZIP file. Minidisc uses such a system, so that a 140-Mbyte Minidisc holds the same amount of audio as a CD containing 720 Mbytes of audio. This uses psycho-acoustic tricks so that it only records what we hear, and not the background material that is hidden by the foreground.

The Microsoft ADPCM (Adaptive Differential Pulse Code Modulation) form of the wave file (it too has the extension .WAV) uses about a quarter of the data by sending information about the differences between samples rather than absolute values. A 1-minute stereo file that starts out at 10 336 kbytes reduces to 2600 kbytes. The quality stands up very well – it is mainly the high frequency transients that are subtly lost. Converting to mono and halving the sample rate would reduce that to a quarter – one-sixteenth of the original size.

Real Audio and Real Media files are designed to be sent to a modem. They can be 'streamed'. Their associated playback software, which can be integrated with Internet browsers, can play the files as they come in rather than having to wait for the entire file to be downloaded (there is a pause while an audio buffer is filled). They can be optimized for the intended delivery modem. Our 10-Mbyte speech file can be delivered as a 118-kbyte mono Real Audio file. The penalty is a phasiness that sounds as though the speech was recorded in an enclosed telephone box. A larger Real Media file of 367 kbytes sounds better.

Cool Edit Pro version 1 was able to save in these formats, but version 2, at present, does not. Neither version has a direct facility for loading them, as they are intended for delivery, not for editing. If you do need to transfer these files into the editor – maybe you are doing an item about the Internet – you can record the output of Real Audio/Media by setting *Cool Edit Pro* to record the output of the sound card handling the Windows output.

Another format is MPEG (Motion Picture Experts Group), which comes in various flavours with extensions of MP1, MP2 and MP3. MP3 has become increasingly used, as it is very efficient in providing low data, good quality sound in stereo. It can handle different sample rates, and can be set to use different bits per second. The 10-Mbyte wave file reduces to 939 kbytes, retaining 44-kHz sampling and using 120 kbits per second encoding. This gives good quality stereo for an end user. Reducing the bit rate to 48 kbits/s and halving the sampling rate reduces the audio file to 353 kbytes. The quality is still comparable to the original, provided you are listening on PC style speakers.

Cool Edit's MP3 coder can record down to 20 kbits/s. It includes the MP3Pro variant, which is backwards compatible and can be played by ordinary MP3 decoders. However, an MP3Pro decoder can take advantage of extra low bit-rate information that allows it to add back high frequency information. This extra information is a bit of a 'guess', but can substantially improve the perceived quality.

Also provided by *Cool Edit* is Microsoft's WMA (Windows Media Audio) format, which can be used for streaming. It can provide files of similar length to MP3, but its advocates say that it can produce better quality at similar bit-rates to MP3 or Real Audio and Real Media. Microsoft claims that WMA can achieve similar quality at up half the bit rate of an equivalent MP3 file.

Cool Edit is able to save many of the Microsoft audio formats using the encoders that your windows system has acquired either from the Windows installation disk or when downloading other programmes like the Windows Media Player. These are using the ACM waveform option in *Cool Edit*'s Save dialog. The options button then gives you access to whatever MS formats you have on your machine (Figure 14.4).

All these compression formats make assumptions about your ears and the programme material. In practice, some music and some voices compress better than others. For example, a light male voice often sounds less 'phasey' than a deeper male voice at high compression ratios.

Whatever your final format, it is vital that the sound file is properly prepared. Do everything at full 16-bit resolution and your normal sampling rate. Only when everything is ready should you reduce the sound to the lossy compression format.

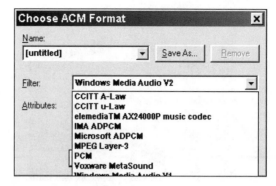

You may wish to try adding a little more mid- to high-frequency presence to the sound to help it punch through. Use the hard limiter to lose isolated peaks. Maybe a little audio compression will help. Save the audio to your chosen format and listen to it critically. Repeat until you are satisfied.

Figure 14.4 Using the ACM waveform option

Beware: when you SAVE AS to, say, an MP3 format, *Cool Edit Pro*'s filename extension will change at the top of the screen, but its buffer will still contain full resolution, uncompressed audio. You have to close the file and reload to hear the result.

Cool Edit Pro can load and save a large number of different formats. Some are obsolete but still found, whereas others are used by different types of computer (such as AIFF for Apple Macintosh computers). The exact formats that you can load will combine those supported directly by *Cool Edit* plus those supported by Windows and any added by your sound card software.

14.9 Analytical functions

Digital audio files consist of a large quantity of numbers. As such they can be analysed statistically and produce at worst pretty pictures, and at best useful diagnostic information about an audio problem. This may be particularly valuable for older users who are losing their high frequency hearing, allowing them to *see* problems that others can *hear*.

Spectral view

Normally when using *Cool Edit* the waveform view is used. This shows how the amplitude of the signal varies with time. However, there is an alternative view; the spectral view. This analyses the spectral content of the signal. Much of the time this is good for making a pretty picture on your screen and little else. However, it can be helpful in tracking down audio problems. Spectral view displays a waveform by its frequency components, where the height represents the frequency and, as usual, time is represented horizontally. This will show you which frequencies are most prevalent in your waveform. The greater a signal's amplitude component within a specific frequency range, the brighter the displayed colour will be. Colours range from dark blue (little audio in this range) to bright yellow (frequencies in this range are high in level). You can change the colours in the Option/ Settings dialog/Spectral 'Colors' tab.

As a diagnostic tool it will show you frequency response problems such as a suck-out at particular frequencies (perhaps resulting from a head alignment problem in an analogue original). It will also

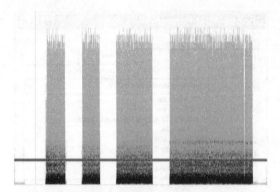

Figure 14.5 Spectral display showing low-level 3 kHz

show the presence of low-level tones (see Figure 14.5) and help you identify their frequency. Know the frequency, and you are half way to finding the source of the problem. It can also be very useful in showing up clicks in audio from transferred gramophone records.

Frequency analysis

The Frequency analysis window (Figure 14.6) contains a graph of the frequencies at the insertion point (yellow arrow cursor) or at the centre of a selection. This window 'floats', meaning that you can click in the waveform on the main *Cool Edit Pro* window to update the analysis while the Frequency analysis window is on top.

The information in this dialog is like one 'slice' or line in the spectral view of the waveform. The most prominent frequency is interpolated and displayed in a window below. You can move the mouse over the graph area to display the frequency and amplitude components of that frequency. Where there is a single instrument playing, you can often see its harmonics quite clearly.

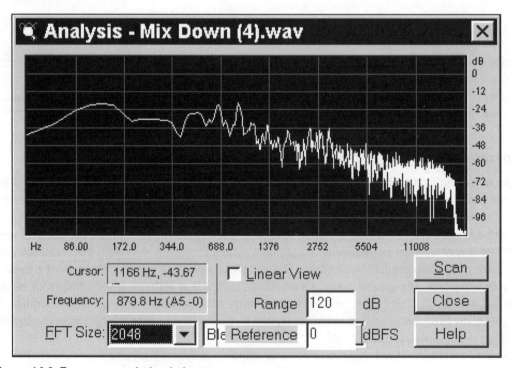

Figure 14.6 Frequency analysis window

You can control how finely the frequencies are analysed. Provided this is not too rigorous for your computer, there will be enough processing power for the graph continuously to update itself while the file is being played. This dynamic display can often reveal low-level (or high-level) tones, etc., that cannot be spotted otherwise. For example, high frequency noise may be present where there is no audio activity. This should be filtered for the benefit of people who can hear it. Similarly, a recording made from FM radio may show signs of breakthrough from the 19-kHz tone that synchronizes the stereo system.

You can also generate a step-by-step animation by clicking on the main waveform window and then holding down on the right arrow key. As the cursor scrolls across the display, *Cool Edit Pro* displays the spectral information in the Analysis window. When you view stereo data, the left and right channels are shown in different colours.

Phase analysis

What is phase?

At its simplest, this is the direction in which your loudspeakers move when a positive audio cycle is applied to them. Do they move forward or do they move away from the listener? The absolute phase – does this correspond with the direction the original microphone's diaphragm moved? – is not thought to be important, although some people dispute this. It is often surprising just how asymmetrical audio can be, especially speech (Figure 14.7). So it is possible that absolute phase could be audible, at least to some people.

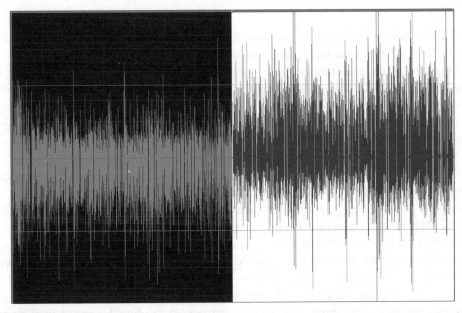

Figure 14.7 Asymmetrical speech. The highlighted area has been inverted to show that this particular speech recording (single male voice reading a story) is biased in one direction even though there is no DC offset

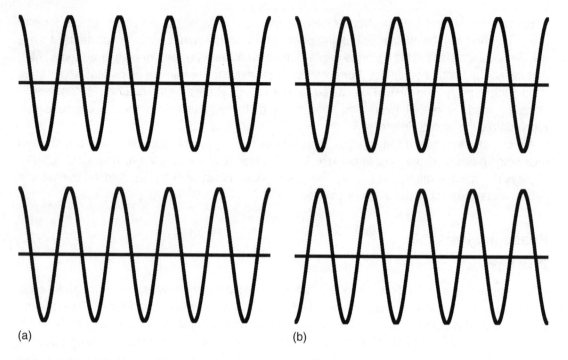

Figure 14.8 (a) In-phase sine waves; (b) Out-of phase sine waves

Figure 14.8a shows a pair of in-phase waveforms; Figure 14.8b shows a pair of out-of-phase waveforms, as the bottom waveform has been inverted.

What is undisputedly important is that a group of loudspeakers must all have the *same* phase. In other words, in a stereo set up the two loudspeakers move together when fed with an identical signal (mono panned centre). If they move in opposite directions this will blur the stereo image and also reduce the bass response, as the two speakers will tend to cancel out low frequencies.

You can tell if your speakers are in-phase by feeding a centre mono signal to them. Do not use tone, which is hard to locate, but choose speech or music with lots of transients – a piano is ideal. If the phase is right then the sound will come from the centre, half way between the speakers. If the phase is reversed, then it will be hard to place the image. On a very good monitoring set-up, the sound will sound as if it is inside your head. This is a good way of assessing how good the monitoring environment is when you are recording in an unfamiliar room. Out-of-phase mono sound can be quite painful – as if it is boring a hole in your head – with good monitoring. If it is difficult to tell the difference, then the room acoustics are poor. In this case, temporarily move the speakers so that they are close together and side by side. Out-of-phase will show itself by a loss of bass.

Professional monitoring equipment often has an easy way of switching the phase of one leg of a stereo signal, but much semi-pro equipment does not, and most computer sound cards do not. It's well worth making a test file that can be played through the system. While you are at it, a channel identification can be added (see 'Making a phase check test file' on page 187).

Making a phase check test file

- Record in mono on to a stereo wave file so that left and right channels are identical. Record words similar to:

   Phase and channel identification . . .
   Identifying left channel only . . .
   Identifying right channel only
   Identifying centre mono both channels in-phase . . .
   Identifying centre mono right channel out-of-phase . . .

- In *Cool Edit*, select the right-hand channel of the phrase identifying the left. Press Delete. You now have that phrase on the left-hand channel only.
- Select the left-hand channel of the phrase identifying the right. Press Delete. You now have that phrase on the right-hand channel only.
- Leave the phrase identifying mono centre alone.
- Select the right-hand channel of the phrase identifying out-of-phase. Click on the menu item Effects/invert.
- Save the file.
- You can now play this file and check your loudspeakers. If you burn it to CD, you can use this to check playback equipment and hi-fi equipment.
- If you find that the phase is wrong – you get a better image with the phase reversed sound – then reverse the connect to *one* loudspeaker only. It does not matter which you change.

Cool Edit phase analysis

Cool Edit implements a simulation of a longstanding hardware way of analysing phase. This was to feed the left and right channels of a stereo signal to the X and Y inputs of an oscilloscope. The X input moves the spot on the screen left and right, whereas the Y input moves it up and down. This means that if there is an X input only, a horizontal line is produced; if there is a Y input only, a vertical line is produced. Mono produces a line at 45° (Figure 14.9(c)). A stereo signal produces a complex image that spreads around the screen.

Cool Edit Pro also has an MS option. What this does is feed the sum of the left and right signals to the Y input and the difference between the channels to the X input. This has the effect of rotating the display through 45°. A mono signal gives a vertical line, a left-hand only signal gives a 45° line going up to the left. A right-hand only signal gives a 45° line going up to the right. This means that the display aligns visually with what you hear, and this is the option that I prefer (Figures 14.9(a)–(d)).

Interpreting the display

If you are copying analogue tapes or cassette, azimuth error is a common problem. Normally the record and replay heads are set to be exactly at 90° to the travel of the tape. If one or the other is

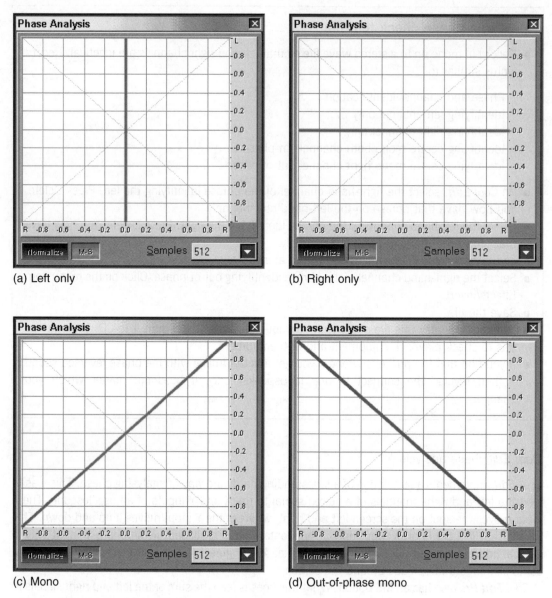

(a) Left only   (b) Right only

(c) Mono   (d) Out-of-phase mono

Figure 14.9 (a)–(d) *Cool Edit* phase analysis (MS option not selected)

out of alignment, then high frequency losses will occur. This can be corrected by realigning the playback head to the setting used by the recording head. This is why it is always advised that cassette recordings are transferred using the same machine that recorded the take, as most use the same head for record and playback.

The reason that high frequencies are lost is that the playback head's 'gap' needs to be of the order of half the wavelength of the highest frequency to be played back. If it is slanting with respect to

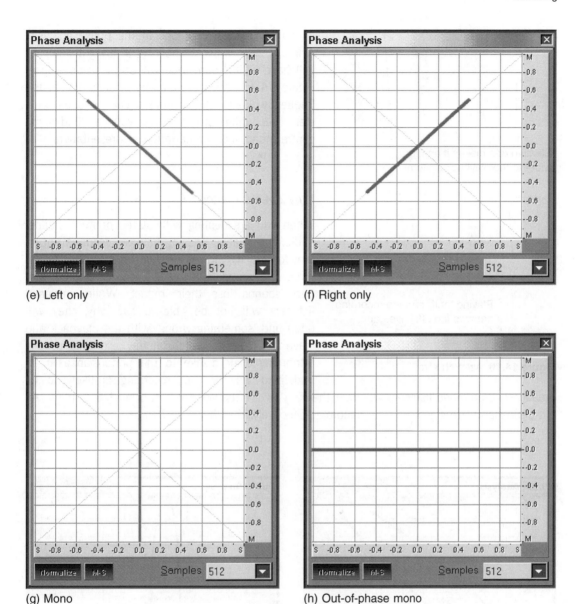

(e) Left only

(f) Right only

(g) Mono

(h) Out-of-phase mono

Figure 14.9 (e)–(h) *Cool Edit* phase analysis (MS option selected)

the recording' then the top of the gap is scanning a different part of the recorded waveform to the bottom. This makes the gap appear as if it were wider than it really is.

A quick fix for a mono recording is to play it back with a stereo head and only use one track. This halves the effective gap width error (see Figure 14.10). Adjusting the head for best high frequency output by ear is harder than you might think. However, the *Cool Edit Pro* phase analyser can give you an objective view of this. Play the mono recording through a stereo

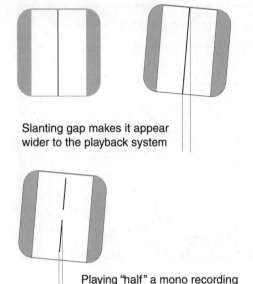

Slanting gap makes it appear
wider to the playback system

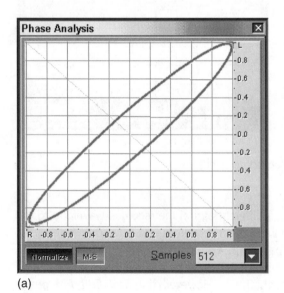

Playing "half" a mono recording
narrows the effective gap

Figure 14.10 Azimuth error

machine as usual and look at the phase analysis. Azimuth error will not show as a straight line, but will have width; pure tone will show as an ellipse. Figure 14.11 shows a dramatic error of 15° at the frequency used. This technique can be used on a stereo recording, provided that you can find some centre mono such as narration or presentation.

Why does it matter?

When you are dealing with stereo or multichannel audio, the relative phasing between feeds is important for exactly the reasons we used to identify the error. The soundstage image becomes blurred, and bass sounds lose their impact. While untrained listeners will not be able to say why, they will often find 'something wrong' with the playback and leave the room or switch off the recording, usually attributing their dislike to the performance in an undefined way. If your recording is likely to be broadcast, then mono listeners will hear the sum signal (the vertical part of an MS display). An out-of-phase centre signal will be inaudible. Effectively, they hear the *Karaoke* version!

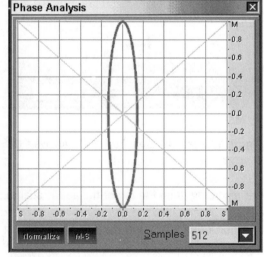

(a)

(b)

Figure 14.11 A 15° phase error; (a) LR; (b) MS

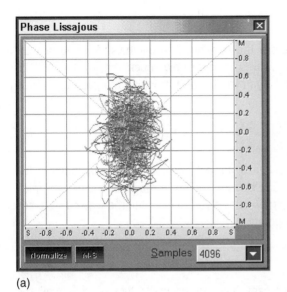

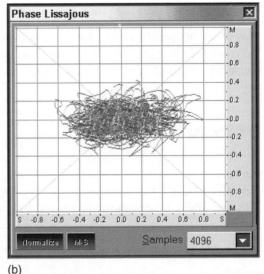

(a)   (b)

Figure 14.12 MS display: (a) In-phase stereo audio; (b) Out-of-phase stereo audio

You can very quickly develop 'an eye' for using the display to check your recordings for phase errors, as on average the sum signal will be slightly larger than the difference. Figure 14.12 shows an identical stereo signal in-phase and out-of-phase. The Normalize button amplifies the display so that it is visible even on low-level displays. The samples dropdown controls the number of samples used; the more samples used, the denser the display.

Statistics

Analyse/statistics (Figure 14.13) produces information about the whole waveform, or the section that is selected.

The Statistics dialog can be used to get the following details about the current waveform.

Minimum/maximum sample value

Minimum, in this context, does not represent the 'quietest sample'. It couldn't, as every cycle of audio goes through zero. What it does represent is the peak negative-going excursion of the waveform. In both cases, clicking the arrow will take you to where this maximum or minimum is.

Figure 14.13 Waveform statistics

Peak amplitude

This combines negative and positive, and expresses the peak level in decibels. A value of 0 dB indicates a fully modulated wave file. Negative values indicate how much the level would be increased if normalized. Clicking the arrow will take you to the point at which the peak amplitude occurs.

Possibly clipped samples

A sample that has the value of the highest possible or lowest possible value can indicate full modulation. However, because no higher number can be represented, it may indicate that clipping – digital overload – has taken place. Clicking the arrow allows you to examine the waveform to see if there really is a problem, by taking you to the point.

DC offset

This reports the DC offset to within 0.001 per cent, and represents a test of how well the manufacturer has set up your sound card. Each sound card input may have a different value. Ideally, they should all be zero.

Minimum/maximum/average/total RMS power

This is roughly equivalent to the area of the waveform display. A very peaky waveform may consistently touch 0 dB but it may have less power than a continuous sound of lower level. The arrows will take you to minimum and maximum points, which may well be in different places from the minimum and maximum amplitudes.

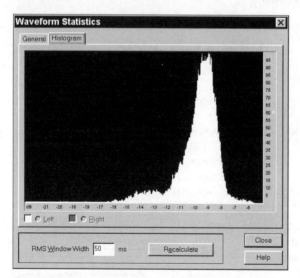

Figure 14.14 Histogram tab

Histogram tab

The histogram tab (Figure 14.14) provides a way of seeing the distribution of amplitudes in your recording. The graph shows which amplitudes are most common. The horizontal ruler measures the amplitude in dB, while the vertical ruler measures percentage. The more audio there is at a particular amplitude, the more the graph will show for that amplitude.

15

CD burning

15.1 Types of CD

The audio CD, designed in the late 1970s, is now quite an old format. Today, its specification quirks can be quite frustrating. The audio CD was never designed as a filing system, and fetching data to bit accuracy is not straightforward. CD-ROM adds the necessary file system information, but at the expense of reducing data capacity. However, it remains a robust format and has spawned a large number of other formats using the same sized optical disk.

The audio CD is what we would now call a streaming format. It was designed to provide a continuous bit stream to a CD player. It was also designed to allow errors. It is not a filing system; that rôle is taken by the CD-ROM. This can be confusing if you are new to making your own CDs. You can make either from the same blanks. They look the same, but one will be played by a CD player and the other spat out, played silently or, at worst, played with tweeter-searing loud noise that in no way resembles your audio.

It all comes down to error correction. A 74-minute CD blank can record 720 Mbytes of audio but just 650 Mbytes of data. The audio tracks are allowed to have errors, which the CD player will correct sometimes, but more often than not it makes a guess at the missing value (interpolation). Otherwise, it just mutes what might well be a loud click. The CD-ROM uses the 'missing data' to apply more robust error correction techniques. For more information on error correction, see Chapter 2.

The CD-ROM can contain a very large number of files. These can be any type of data, including audio of any format. For example, using very low bit rates you can get 500 or so MP3 music files onto a single disc. CD audio discs are much more restricted. They have an absolute maximum of 99 tracks, and each track must be at least 4 seconds long. The first track should start 2 seconds in to the disc. The tracks must be stereo and use a 44 100 kHz sampling rate with 16-bit resolution.

Each track can have up to 99 index marks, the first one being the beginning of the track. The original idea was that a track would contain a single classical work, with each movement marked by index marks. Although this is part of the original CD specification, many CD players do not make use of this facility. The corollary of this is that not many commercial CDs handle them either. Even so they are used by many sound effects collections, which cram many more than 99 sounds onto each CD – for example, a track of 20 door slams would have each slam marked with an index mark.

15.2 Audio CDs

Because of the streaming nature of CD audio, the tracks do not have file names and were not designed so that points within the track could be found with bit-accuracy. The first has been addressed partially by an addition to the CD specification that allows text to be associated with each disc and each track so that information can be added, rather like on a Minidisc. Again, many CD players ignore this information.

The second has made CD copying and recording problematical. The 'easy' way of copying a CD is merely to play it and use the CD-ROM drive's audio output to record. This can only be done in real time and includes the limitation of the analogue section of your sound card. Reading the CD digitally has to be done in blocks, but when the CD-ROM drives goes back for the next block it cannot find where it finished to sample accuracy. This can lead to what is often called 'jitter', as the resulting files can have bits of audio missing along with sections where there is an overlap. The normal way of dealing with this is for the software deliberately to fetch overlapping blocks and then line them up to find identical sections and electronically splice the blocks together on-the-fly. This software can be in the computer or contained within the CD-ROM's firmware.

Not being able to find exact points within a CD track also causes the traditional problem with recording CDs. CD burning is not something that can usually be interrupted, and this means that many computers cannot be used for anything else while the CD is being made. However, the good news is that the CD can be burnt in less than real time. Computer CD recorder drives that can handle ×12 and faster are common. In itself, a fast drive is not enough; the computer system itself must be fast enough to be able to provide the data without a break.

Everything is fine provided a continuous bit stream can be sent to the recorder. However, computers can be doing other things, and there may be a delay before the next block of data is sent. To counter this, CD recorders have memory buffers of several megabytes so that they can continue to work on the buffered data during any gaps from the computer. As you can record CDs at much faster than real time, even quite large buffers of several megabytes are often not large enough. While a buffer may be able to contain the equivalent of 40 seconds of audio, if you are burning the CD at 16 times real time then it is only proof against a break in data of $2\frac{1}{2}$ seconds. Running out of data in the buffer is known as a 'buffer under run'.

With older CD burners the only option is to ensure that the computer is doing as little as possible in the background, with software such as virus checkers switched off. Ideally the audio should be on a hard disk that has been freshly defragmented to speed up the supply of data. Over time, data files are fragmented around the surface of the hard disk. This means that the head has to dodge about collecting the data. Unfragmented files allow the head to run continuously from beginning to end of the file. The Windows software utility to defragment files is usually found in the Start menu under: Programs/Accessories/System Tools/Disk Defragmenter.

The program itself is C:\WINDOWS\DEFRAG.EXE. If you do not have this, then either install it from the Windows CD or use a third party defragmenter. Routine defragmentation of files is good practice anyway, and helps prevent your computer slowing down over a period of weeks.

If the buffer empties, even only momentarily, then the CD recording process fails and you are left with a useless disc – often dismissively referred to as a 'coaster' or 'table mat', as it is good only for protecting your desk from your cup of coffee. However, recorder speeds of up to ×4 should be well within the capabilities of all but the most ancient of machines.

Fortunately technology has come to the rescue and CD recording drives are coming with improved techniques of dealing with this, such as Burnproof® technology. With this, the CD burner monitors the input buffer. If it falls below a certain capacity, say 10 per cent, it stops the recording process in an orderly way. Once the buffer has filled again, it can match the data in the buffer against the last block it recorded, play back what it has already recorded, then drop in to record and continue without a gap. Actually there will be a tiny gap, but it is within the tolerance specified from the very beginning of CD manufacture and so all CD players should have no problem. This technology also works for data being recorded as a CD-ROM.

15.3 Multi-session CDs

Most CD blanks, CDRs, can only be written to once – there is no way of erasing them and starting again. There is a halfway house where you can make 'multi-session' CDs, which allows you to add tracks one at a time in separate sessions. In order to achieve this, the table of contents for the CD is written in a different place. The effect of this is that most CD players will not recognize the disc while most computer CD-ROM drives will. The capacity of the disc is reduced, as there is recording space lost each time for internal housekeeping. When you have finished adding tracks you can 'fix' or 'close' the disk, and what this does is to copy the latest table of contents to the place on the disk where ordinary CD players expect to see it. At this point no more tracks can be added, but ordinary CD players can play the tracks. This made a lot of sense when the cost of a CD blank was measured in tens of pounds or dollars, but nowadays, with blanks being so cheap, the ability to create multi-session CDs is losing its usefulness.

15.4 Rewritable CDs

Rewritable and erasable CD blanks, CD-RWs, are easily available at quite low prices. At first glance, these seem to be a panacea. However, they are still more expensive than CDR and have several disadvantages – the major one being that they cannot be played by most ordinary CD players. They have a lower reflectivity compared with standard pressed CDs, and need a stronger laser to read them. Many DVD players can read them as a by-product of being able to read DVDs.

CD-RWs have to be erased before reuse, although this usually only means wiping the table of contents and is quite quick. You cannot erase a single track as such. Using CD-RWs as 'large floppies' is a very attractive idea, and there is software that allows you to do this. How successfully it does this seems to be very machine-dependent – some people have used CD-RWs in this way with no problems at all, while others have had nothing but 'blue screens of death' as their computers crash. The capacity of the disc is much reduced by the need to format it, which takes time, and a large amount of space is taken by the formatting information and cannot be used to record. Quite often other machines using nominally the same software cannot read the discs, so the portability of CD is lost.

As with multi-session CDs, many people take the view that CDRs are so cheap that the extra hassle of CD-RW is not worth it. Others compromise and just use them written as if they were CDRs, only making use of their erasability.

15.5 CD recording software

Normally basic CD software requires you to specify if you are burning an audio CD, a CD-ROM or a mixed format. Tracks are dragged into a window and the CD burnt. If you are planning to produce actual audio CDs, then more sophisticated software is needed. Simple software works on the basis of one wave file per track.

This need not necessarily be so. Unfortunately no longer available, *Sound Forge*'s CD Architect could use one or more wave files, and it was up to you how they were spaced and where the track markers were (Figure 15.1).

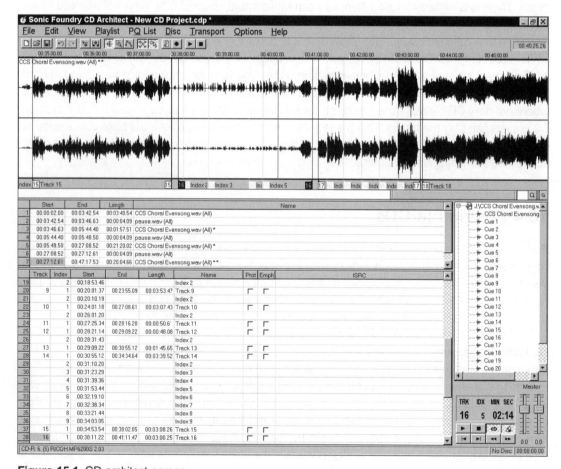

Figure 15.1 CD architect screen

For CDs of events, you can vary the length of each inter-track gap or have none at all. You can even continue the audio through the inter-track gaps. For a CD of a concert you can arrange for each song to begin a track, but for speech links and so on to be in the gaps between the tracks (the CD player counts down the seconds in the gap). The CD will play through continuously, but a track skip will take you directly to the next song.

If you have access to professional or semi-professional CD players that will make use of them, you can also add index marks. These can be useful to you for archival purposes, allowing you to mark individual items without starting a new track. You can mark individual contributions within a feature or program. Unlike tracks, index marks do not appear in the table of contents and the CD player literally has to scan through the track to find them, so index searches are slower than track searches.

Traditionally, CD burning software offers various options as to how the disk is recorded. You decide whether you want to record the whole disc at once or to go for the multi-session option. If you do, you often have to remember to 'Import' the last session so that the new recording is added to the previous table of contents.

Cool Edit Pro CD software

Version 2 of *Cool Edit* provides audio-only CD burning that can insert indexes as well as tracks. A single wave file can be split into several CD tracks, not necessarily in their order in the wave file.

The CD burner is integrated into CEP's 'engine' as a plug-in. The first task is to select your tracks by loading them into the wave editor. Clicking on 'File/CD Burning' will bring up a dialog with your files in the left-hand window (Figure 15.2).

These can be dragged across to the right-hand window to select them to be burnt to CD. The size of the dialog can be changed by dragging with the mouse in the usual Windows way, as can the

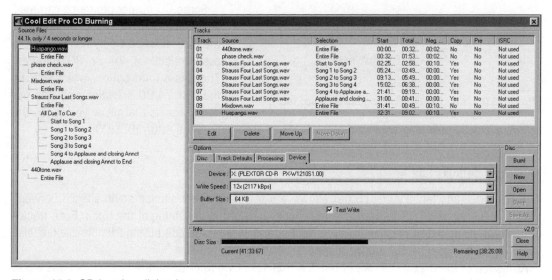

Figure 15.2 CD burning dialog in use

width of the columns. Files shorter than 4 seconds are ignored, as tracks must be 4 seconds or longer to conform to the CD specification. As you will see in the illustration, some files have subsections that can be dragged separately so that parts of the file can appear as separate tracks, with or without silences between them. There is also an option to drag 'All cue to cue', which will load all the cues in order with the first one being 'start' and the last 'end'. *Cool Edit* can handle CD tracks marked as regions or as points. The file uses a different marker from what are now described as 'Basic' markers, and the track marker is created using Shift/F8. This will create a track that corresponds to the current selection or a track marker at the current cursor if there is no selection.

Both types of marker have their uses. If you have a continuous recording of, say, a live performance, then all you may want to do is to mark particular events in the recording so that listeners can find their way around easily. This is easily done by marking a track at the current cursor position. However, it may be that you want to extract clips from a recording. Mark them as regions by track-marking a selection, then these clips will become available as separate tracks. There is also an 'All Regions' option. If this is dragged to the tracks list, then all your clips will be copied across automatically. You can mix regions and cue points in the same wave file.

Within the dialog there are four tabbed pages: the disc tab, the track defaults tab, the processing tab and the device tab.

The Disc tab

The Disc tab (Figure 15.3) allows you to select two options. The tick box allows you to give the CD a catalogue number, and theoretically this should conform to the UPC/EAN standards, sometimes known as bar codes. However, you can utilize your own numbering system for personal use if you wish.

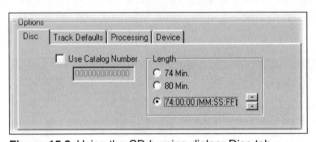

Figure 15.3 Using the CD burning dialog: Disc tab

Some CD burner programs can read the available duration information off the disc. *Cool Edit* only has selectable assumptions of 74 or 80 minutes. There is a third option, which follows the last 74/80 selection and has up/down arrows that allow you to tell the software exactly how long your disc blank is. They are rarely exactly 80m:00s:00f.

The Track defaults tab

The Track defaults tab (Figure 15.4a) allows you to select either silence or no silence between tracks for the whole CD. The silence is technically part of the beginning of the track. Each track's options can be edited by clicking on it to select it, and clicking the edit button (double-clicking also works). This brings up a sub-dialog (Figure 15.4b).

At the top of the dialog is a pair of buttons that select whether default or custom options are to be used. Clicking 'Use Custom Options' ungreys the rest of the dialog:

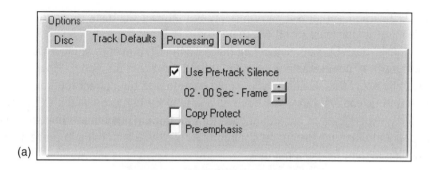

(a)

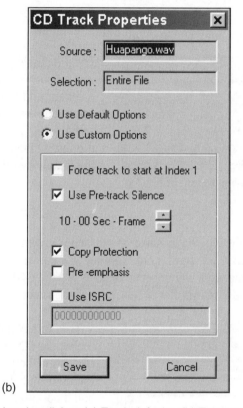

(b)

Figure 15.4 Using the CD burning dialog: (a) Track defaults; (b) Track properties editing dialog

- The 'Force Track to start at Index 1' option forces the track to begin at the first index point. This allows the user to add indexes or pre-track silence without specifically creating a negative area.
- 'Use Pre-track Silence' allows you to select the duration of the silence; the arrows change the value in single frames. The default is 2 seconds.
- The 'Copy Protect' box will set the copy protect flag on the resulting CD's datastream. This is a very polite form of copy protection. It merely indicates to another device that cares to read it 'Please do not copy me'. Many devices, especially computer sound cards, take no notice of this

request. Nevertheless, it remains a way of asserting that your CD track is copyright and should not be copied, but it is a cruel world out there.

- The 'Pre-emphasis' box, if ticked, sets a flag that tells the CD player to apply a standardized top cut to the playback to compensate for a standardized top boost on record. However, *Cool Edit* does not actually apply this itself and you would have to add the correct top boost to your file. This is little used nowadays, but is intended as a way of further improving a CD's noise level.
- 'Use ISRC' allows you to set an individual code for that track. Theoretically this should conform to the International Standard Recording Code format, but there is nothing to prevent you having your own coding system for personal use.

The Processing tab

The Processing tab (Figure 15.5) allows you to analyse all the selected tracks for loudness and then process them to be the same loudness. This may well not be the same level, as it can take into account the ear's different sensitivities throughout the frequency range.

Figure 15.5 Using the CD burning dialog: Processing

The Device tab

The device tab (Figure 15.6) is where you select the device you want to record onto. Most people will only have one burner, but if you have more than one you can select between them here. The write speed set at 'Max.' should work with most set-ups, as will the 64-kB buffer setting. If you have problems, set a different value for the buffer or select one of the slower speeds that your burner can cope with.

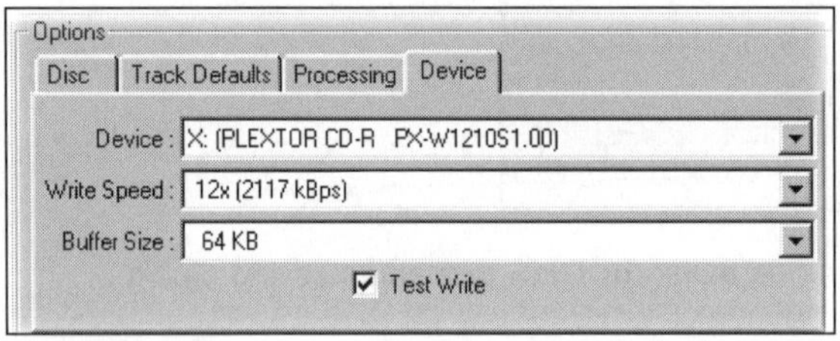

The 'Test Write' box will, if ticked, cause the whole CD burning process to be gone through without actually burning a CD. It's there so that you can check your system and if necessary alter it, without producing a lot of duff discs. Once you know your system, you are unlikely to use this very often.

Figure 15.6 Using the CD burning dialog: Device

Creating CD tracks from within files is done by using *Cool Edit*'s cue markers. These can be set so that the CD burner sees them as potential tracks. In the source file list (Figure 15.7) they are shown inset underneath the complete file. This means that you can drag the complete file to be a CD track, or one of the selections within the track. Tracks can be selected by selecting a region within the file, but instead of just pressing F8 to mark a region use Shift/F8, which will create a track. This will be named by *Cool Edit* 'Cue 1', 'Cue 2' . . . etc. In the example shown, 'Cue 1' for 'Sailing By.wav' is missing because, as you can see from that file's cue list (Figure 15.8), it is less than 4 seconds.

Additionally, the third track 'Cue 3' has had an index mark inserted. You do not have to accept *Cool Edit*'s default names; they can be edited in the adjacent box in the dialog. In this case it has been renamed 'Cue 3 and a bit'. Indexes are inserted at the cursor position using Control/F8. If you make a mistake, then you can edit the markers. A positional error is fixed either by dragging the cue marker within the file or by enabling the editor within the cue list dialog and typing in values. The type of marker can be changed using the drop down shown at the right (Figure 15.9). The editor is revealed by clicking the 'Edit Cue Info' button. Depending on how the cue list window is sized, the edit panel will appear at the side or the bottom.

Burning the CD

As you add your tracks the disc size bar (Figure 15.2 above) will grow bigger, telling you what duration of audio you have added and what time remains on the disc.

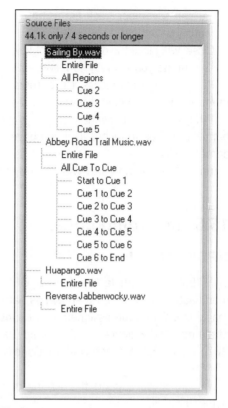

Figure 15.7 Example of source files display

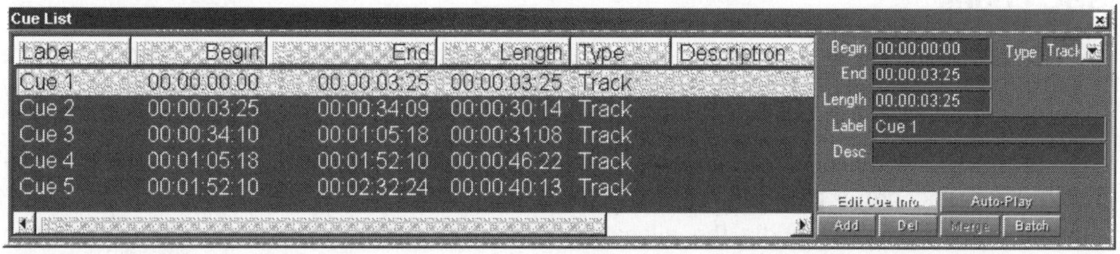

Figure 15.8 Cue list showing Cue 1 less than 4 seconds and index marker in Cue 3.duration

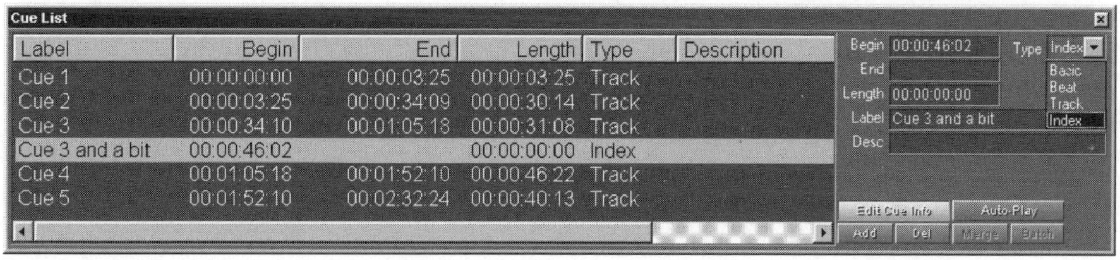

Figure 15.9 Cue list illustrating index marker with edit panel enabled showing marker type dropdown

You can save your track list using the 'Save' or 'Save as' buttons. 'Open' allows you to load a previously saved list.

'Close' will close the dialog but not lose your list. 'Burn!' brings up a confirmation alert from which 'OK' commits you actually to record onto your CD blank.

At the end of the process, CD burners will eject the disc. This is so that Windows will be forced to reread the disc when it is reinserted and recognize that it has changed.

If the disc is for someone else, you may want to check it before sending it off. Don't use the CD burner you used to record it! If you have a separate CD-ROM then use that, or, even better, use a domestic CD player. If you are paranoid about the quality, you can rip the newly created CD and compare the resulting files with the originals.

15.6 Recording codes

UPC/EAN code

EAN (European Article Number) or UPC (Universal Product Code) codes identify products and manufacturers. Sometimes known as bar codes, they are usually only required for professional CD mastering. Codes are assigned on a country-by-country basis. In the UK, contact: Association for Standards and Practices in Electronic Trade, EAN UK Ltd, 10 Maltravers Street, London WC2R 3BX. In the USA, contact: Uniform Code Council, Inc. 8163 Old Yankee Street, Suite J, Dayton, OH 45458.

Not all drives support the writing of bar codes. Consult your CD-R drive documentation to determine this.

ISRC codes

The tracks can have individual codes. They use another system called Industry Standard Recording Codes (ISRC). These are designed to provide an easy way to log the play of specific CD tracks by broadcasters.

In the UK, for further information about the ISRC system please contact: PPL; http://www.ppluk.com or telephone 020 7534 1122.

In the USA the RIAA handle the allocation and have more information on their web site (http://www.riaa.org/Audio-Standards-3.cfm) including a description of how the code is formatted. Alternatively, telephone 1–202 775–0101.

Not all drives support the writing of ISRC codes. Consult your CD-R drive documentation to determine this.

16

Archiving

16.1 Selection for archiving

Archiving often involves difficult decisions. If you keep everything, then you soon will be buried in old recordings. To be any use they need to be catalogued, and this is a lot of work. Yet so often the interview that is thrown away is the one that turns out to have commercial value because of later fame or notoriety acquired by the person interviewed.

16.2 Compact disc

The ease with which computers can burn CDs has simplified the matter. A CD blank can contain about 80 minutes of audio in standard CD audio format. The discs themselves take up relatively little space, especially if kept in folders rather than conventional 'jewel cases'. Plastic sleeves punched for insertion into standard lever arch folders allow the discs to be kept with any paper documentation.

Keeping your material as CD audio means that it is easily accessed using ordinary compact disc players. The audio can be reloaded into the computer faster than real time (×12 or more). However, editing information is lost and there can be a small loss of quality for each generation of copying due to the compromise nature of CD audio recording, which was never designed as a data storage format.

The Track At Once option of much CD burning software has the advantage that it allows you to burn multi-session CDs. This could be useful for archiving, as you can add a track and remove the CD. On another day you replace the CD and add another track. While you are doing this, the CD will be readable on your computer but *not* on an ordinary CD player. When you have finished adding tracks the CD is 'closed' or 'fixed'; no more tracks can be added, but the disc now becomes readable on an ordinary CD player. However, adding a track to a multi-session CD can fail and you are likely then to lose the entire contents of the CD.

A better strategy for archiving is to allocate hard disk space for your archive material and then burn the CD in one go. Make two copies for real security. Ideally, archived material should be kept as computer files in CD-ROM form, as this is error free.

16.3 CD-ROM

CD-ROM is designed as a data storage format at the expense of a little storage capacity – a little over 700 Mbytes of data for an 80-minute CDR blank. For most medium-length items this is enough to contain all the material used in the editing session, including the edit decision lists (session files in *Cool Edit* speak). This means that you can access the complete editor information and can 'unpick' edits or access the original uncut material. The data CD can also contain word processor and text files, and even pictures associated with the item. This may include transcripts if they were prepared for editorial, copyright, or World Wide Web use.

On the other hand, the convenience of having a CD audio version is very great, so a good compromise is to burn one of each. The audio CD becomes your everyday 'library' copy – you can take it home, or to someone's office, and listen to it on any CD player. The CD-ROM is kept on the shelf and used only to transfer material to be reused. If the CD audio version becomes damaged, a new one can be made from the data files on the CD-ROM version. In extremis a new data CD-ROM of the audio can also be made from the audio version, although this will, of course, have none of the extra non-audio information that was on the original CD-ROM.

16.4 DAT

DAT cassettes can be used for archiving. Audio cassettes have 2 hours of capacity at 44.1 kHz or 48 kHz, and twice that at their half-speed setting. This records at 32 kHz sampling rate. Using DATs manufactured for computer backup, you can get 4 hours or more onto a single cassette at full quality (8 hours or more at LP speed). However, the tape is thinner than audio DAT machines like, and actually using them for computer backup is a much better option.

DAT can be useful as a way of archiving 48-kHz masters in a way that can be played as audio on conventional machines. DAT has a number of disadvantages:

1 The tape is very fragile and is prone to humidity problems. When a DAT gets mangled in a machine you have lost everything. While rescuing it is technically possible, this is a very expensive and time-consuming activity, with no guarantee of success.
2 While it may sound trivial, a substantial disadvantage is the small size of the cassette. There is little room to write an informative label on them, and they have also been known to disappear down the backs of chairs! The CD with jewel case or folder has a definite advantage here.

16.5 Analogue

It is possible to archive material onto analogue media such as compact cassette or reel-to-reel. This has no advantage except, perhaps, short-term compatibility with an existing system. Sooner or later, the existing system will have to be transferred to digital; it makes sense to do it sooner rather than later.

16.6 Computer backup

Just as all your word processor files and web transcripts can be transferred to the same CD-ROM as your audio, so this can be done using whatever computer backup system you have – you do have a computer backup system, don't you?

All computers crash. All computer systems lose data. No one believes in backups until the day they lose a day's, week's, month's, year's work. Companies have gone bankrupt because of poor (or no) backup strategy.

A comprehensive backup strategy will also take into account the possibility of theft and fire. Programs are (usually) replaceable. Data files represent an investment of time and money, and cannot be replaced unless you have a backup.

Even if you do have a backup, do you know how to restore your files? A cynic would say that the world is full of good backup systems that are rubbish at restoring. Beware; while it is easy to protect against the theft of information by password protection and encryption, this protection is lost if the password is written on a Post-it® note stuck to the side of the monitor. Equally, if the computer is lost in a fire then so too will be the Post-it note! You must have a safe place to keep the password.

Choosing a password

Passwords should be easy for you to remember but difficult for others to work out. If you must write a password down, then leave it in a secure place. Avoid the obvious; 'secret' is one of the first things that a malicious hacker will try. They will also try the names of your spouse and any children.

For the best protection, you should use at least six characters and avoid dictionary words. For extra safety include non-alphanumeric characters, such as &$*@%.

With some programs the password may be case-sensitive, so that 'rad$38' is treated as a different password to 'Rad$38'. Here, the CAPS LOCK key being on can cause initial panic when the password you are so confident about is not accepted!

If you are an employee, then your company should have an administrative system where your password(s) can be kept securely against the day when you meet your demise in a car accident or, with long-term data, the company needs to recover the data a few years after you have left it.

Storage systems

There are a plethora of hardware data storage systems, some based on tape cartridges and some based on disk cartridges. Because of the possibility of theft of the computer or of fire, the backup must be on some form of removable medium. Tape cartridge systems have the disadvantage of a relatively short cartridge life. The tape wears, and some recommend cartridges being replaced after anything from 20 to 100 uses (this includes DAT).

With the dramatic fall in price of hard disks, it is worth considering using standard high-capacity hard disk drives mounted in removable caddies for your backup.

Your backup regime should rotate round at least three removable media – the so-called grandfather, father, son strategy. Again, because of fire you should store them away from your computer, in another room or even another building – especially with mature, commercially valuable data. One way of achieving this is to backup, via a network connection, to a remote server. This can use a local area network or, if you have fast enough access, the Internet. For the amount of data involved in audio files you will need at least ISDN or ADSL speeds, and even so this will involve overnight backup runs. There are firms that provide off-site storage on their own computers. Talk to your system administrator about preferred methods of off-site backup.

Despite all this, for many purposes taking a backup home with you overnight can be an effective way of meeting reasonable backup safety criteria.

A related problem is how to move material from one place to another. Sending a CD in the post may be adequate, or an audio file can be 'emailed' from computer to computer. Some radio stations use their administrative computer network to send audio data, often at less than real time, as and when there is spare capacity.

17

Tweaks

Computer programs always have options that allow you to tweak how they work so that you can make them easier to use. *Cool Edit Pro* is no exception.

17.1 Mouse options

Your choice of mouse will make a difference to *Cool Edit Pro*, as it exploits a wheel mouse – the wheel can be used to zoom in and out in the single wave view and to move up and down the tracks of the multitrack view. Within the Effects transform windows you can use the mouse wheel to alter settings. This is done simply by holding the mouse cursor over a slider and rotating the wheel.

Syntrillium have changed the way the right mouse button works in version 2 but have generously catered for those of us who feel that the old way was better. The Options/Settings/General tab has the options for 'Edit View Right Clicks' of the Pop-up Menu or 'Extend selection (Hold Control for pop-up menu)' (Figure 17.1).

The default action is that when you right-click in the wave editor you will get the pop-up shown in Figure 17.1, giving useful options. The downside of this is that the massively useful right click to extend a selection is accessed by having to press control as well. The options in the pop up are available directly from the top menu, so many people prefer to use the original *Cool Edit* option of reversing the functions so that a right click extends the selection and you press control to get the pop-up.

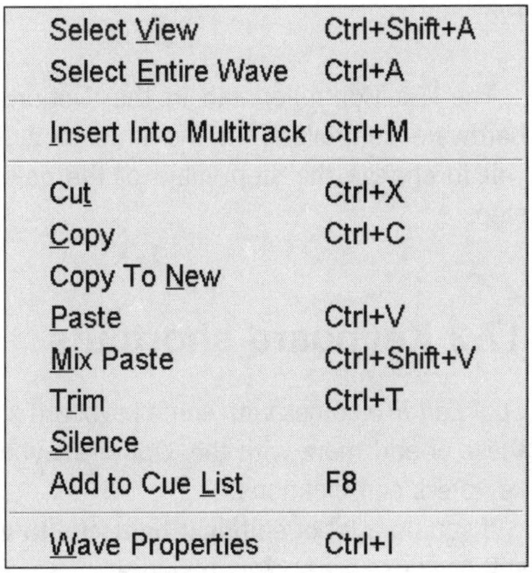

Select View	Ctrl+Shift+A
Select Entire Wave	Ctrl+A
Insert Into Multitrack	Ctrl+M
Cut	Ctrl+X
Copy	Ctrl+C
Copy To New	
Paste	Ctrl+V
Mix Paste	Ctrl+Shift+V
Trim	Ctrl+T
Silence	
Add to Cue List	F8
Wave Properties	Ctrl+I

Figure 17.1 Right click pop-up

17.2 Hardware controller

The disadvantage of all computer-based editors is that you only have one 'finger' – the mouse pointer. With the availability of USB and Firewire ports it is becoming increasingly practical to have an external control surface to provide you with extra controls that are dedicated to your program. Windows 98 onwards provide 'Human Interface' drivers that are allowing a degree of standardization.

Controllers can actually be full-scale mixers with faders and EQ controls, but these are understandable expensive.

Syntrillium provide an optional extra of a dedicated controller for *Cool Edit* called 'Red Rover' (Figure 17.2). This is largely aimed at users who want to record themselves with *Cool Edit*. For

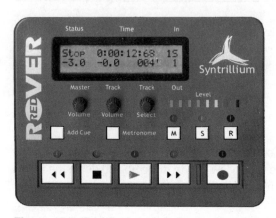

Figure 17.2 'Red Rover' controller for *Cool Edit Pro*

example, a guitarist would have problems with the electric pick up, also picking up the electrical interference from the computer monitor or even the computer itself. Moving a few metres away can eliminate this. Red Rover provides transport controls and level indication, and these can also be useful when editing. Additionally, cue markers can be added on-the-fly and *Cool Edit*'s metronome switched on or off.

In the Multitrack view, it provides an easy direct control of the master volume. Another knob is used to select any single track, and that track's volume can also be changed. Its Mute, Solo and Record enable buttons are available, along with indication of their current settings.

The 'Ext Controller' tab in the 'Options/Device Properties' allows you change the settings of hardware controllers. At the time of writing, Red Rover has just the one option that allows you to change the step value of the gain control knobs between 0.1 dB and 0.5 dB in 0.1 dB steps.

17.3 Keyboard shortcuts

Cool Edit Pro comes with some keyboard shortcuts already defined (Figure 17.3a). You can change these or add more with the 'Options/keyboard dialog'. Set up the actions you use most as simple key press combinations.

If you do a lot of editing it be useful to give yourself some spool controls on your keyboard. The settings I use are shown in Figure 17.3b. This allows me to play a file and then go into continuous fast forward with a single press of the '3' key on the numeric keypad. Hitting '1' reverts me to play; '7' and '9' do the same for reverse wind; and '4' and '6' become momentary reverse and forward spool buttons. I also allocate the '+' and '–' keys on the numeric keypad to operate the vertical zoom controls.

You may also wish to consider reallocating the 'Selection anchor left when playing' and 'Selection anchor right when playing' controls from '[' and ']' to '0' and '.' on the numeric keypad ('.' is shown as 'decimal' on the shortcut list). These keys are easier to find than '[ ]', which sit on little-used keys.

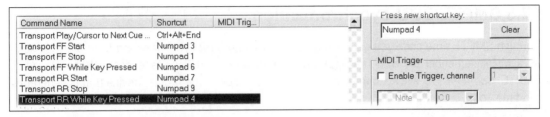

(a)

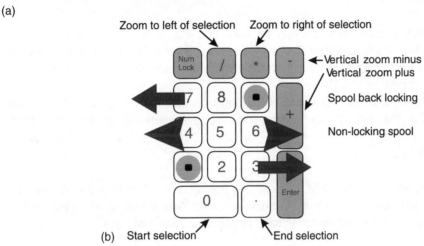

(b)   Start selection   End selection

Figure 17.3 (a) Transport keyboard shortcuts; (b) Suggested layout for numeric keyboard 'spool' controls

17.4 MIDI triggering

Shortcuts can also be set to MIDI commands. These could be from a specialist MIDI controller box, a sequencer like Cubase, or even a MIDI keyboard.

In effect, you can turn *Cool Edit Pro* into a simple Sampler (trigger playback from a MIDI keyboard), or even Save your files with a foot pedal if you really want to.

To set up MIDI triggering:

1 To assign a MIDI event to a shortcut, choose 'Shortcuts (Keyboard & MIDI Triggers)' from the Options menu and select the function that you would like to trigger from MIDI. Tick the Enable Trigger box to the right and set the channel and note, or MIDI controller, and click OK. If you are triggering from a MIDI controller you will need to select the controller ID and the Value that will trigger the event. (Example: the default is controller 64 – the Hold pedal. For on/off controllers like this, a value of 127 will trigger the event in the ON position, 0 the OFF position).

2 To enable MIDI triggering, choose 'MIDI Trigger Enable' from the Options menu. You should now be able to trigger *Cool Edit Pro* from your MIDI device. You can get visual feedback of MIDI activity in the lower left corner of *Cool Edit Pro*'s main window.

17.5 SMPTE/EBU synchronization

Through SMPTE time code you can effectively control *Cool Edit Pro*'s transport from a device such as a MIDI sequencer or, with the appropriate hardware, a VCR or tape deck. When slaved to SMPTE, *Cool Edit Pro* will synchronize with frame accuracy to the master device that is generating the time code.

To setup SMPTE sync:

1 Set a SMPTE offset (a time code location for *Cool Edit Pro* to begin playback/record). In Multitrack View select 'View/Advanced Session Properties'. Enter a SMPTE location where you would like *Cool Edit Pro* to begin playback, in the format hours:minutes:seconds:frames.
2 Tell *Cool Edit Pro* to wait for SMPTE.
3 From the Multitrack view, choose 'Options/SMPTE Slave Enable'. Start playback on the master device, and *Cool Edit Pro* will pick up the time code and give a running update in the SMPTE readout at the lower right of the main window. Playback in *Cool Edit Pro* will then begin at your SMPTE Offset location.

Cool Edit Pro requires about 5 seconds of time code, or pre-roll, to establish synchronization. The readout in the lower left corner of the main window will read 'Synchronizing' when establishing lockup, and 'Playback synchronized' when actually established. (See also Appendix 1.)

17.6 Favorites

The Favorites menu can be built to make a collection of your most commonly used actions. Use Edit Favorites to create, delete, edit, and organize items appearing in the Favorites menu.

Edit Favorites can instantly call up any customized *Cool Edit Pro* Effect Transform or Generate effect, Script, or even many third-party tools. The menu can also contain submenus for easy organization.

Figure 17.4a shows an example of how Favorites can be laid out using sub-menus, and Figure 17.4b shows the edit dialogue that created that menu. The '\' character creates a sub-menu. Further entries in that sub-menu are created by repeating the text up to the '\' and then adding a new name. A menu division is created by using multiple dashes as a name. Thus the
. . .

```
Noise Reduction\CEP\Noise Set
Noise Reduction\CEP\Noise Reduce
Noise Reduction\CEP\------
Noise Reduction\CEP\Declick Set
Noise Reduction\CEP\Declick
```

. . . entries produce the sub-menus illustrated.

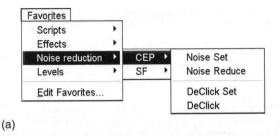

(a)

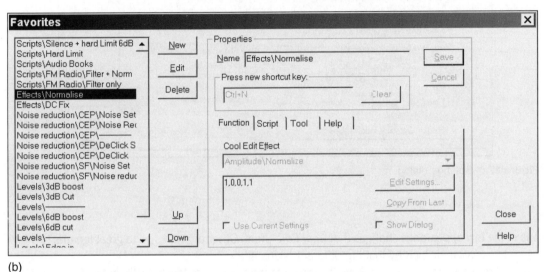

(b)

Figure 17.4 (a) Dialog showing how sub-menus can be created; (b) Edit Favorites dialog

17.7 Settings

The Settings tabbed dialog (Figure 17.5) is the key to customization of *Cool Edit Pro*. This varies from what colours are used by the display to the arcane settings of MIDI and SMPTE parameters. The General tab shown contains most of the settings that can annoy. It is here that the action of the right click in the waveform editor can be changed to the way it worked in version 1. Using the right mouse button to extend the selection is such a frequent event that having to press control is a major slow down. If you have a mouse with a wheel, then the 'Mouse Wheel Zoom Factor' controls how much the window zooms for each turn of the mouse wheel. The default is '80%', but I prefer a less abrupt change of zoom and use a 20% setting.

'Default Selection Range' selects what happens when you double-click on a waveform to select it. The area that is highlighted can be limited to the area you can currently see on screen or select the entire waveform, even if you're only viewing a portion of it.

If it irritates, you can switch off the 'Tip of the Day' that appears when you start *Cool Edit*. It is often useful to have something you have just pasted highlighted, as you will usually want to do more work on it, so ticking 'Highlight after Paste' makes sense. The 'shiny' look is a matter of taste; it's

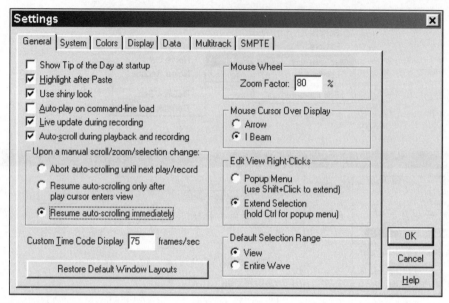

Figure 17.5 Settings dialog

one of the few appearance changes that *Cool Edit* offers. Other programs go to great lengths to allow users to tweak their look by loading different 'skins' files.

Cool Edit can change its layout, as its screen is made up from 'dockable windows'. Figure 17.6 shows the sort of changes that can be made.

The general settings tab has a button, 'Restore Default Window Layouts', that does just that. Version 2.0 of *Cool Edit Pro* does not make any provision for saving different layouts, but maybe later revisions will.

Screen components like the Transport controls, the Time display window, the Level meters, and Rulers may be detached from their current location to float above *Cool Edit Pro*'s main window, or they may be repositioned and resized within the main window so they better suit your requirements.

Dockable windows are indicated by two thin vertical or horizontal lines; these lines are the 'handle' of a dockable window (Figure 17.7). Move your mouse over a handle, and your mouse cursor looks like a plus sign with arrows at each end (Figure 17.8a).

Some docked windows can be resized. If resizing is possible, the docked window will have a single, thicker horizontal or vertical bar, known as a resize control. If your mouse moves over a resize control, your mouse cursor takes on the appearance of two lines with two arrows (Figure 17.8b, c). If you find that despite the mouse changing appearance you cannot resize, then this might be due to other non-resizable windows in the same row. A docked window's right-click menu contains an option called 'Force New Row'. If this item is clicked, a new, empty row to dock windows is created.

To select a docked window, left-click when you're over a handle (the window becomes outlined when selected). To move, select it and drag its hollow shape while holding down the left mouse

(a)

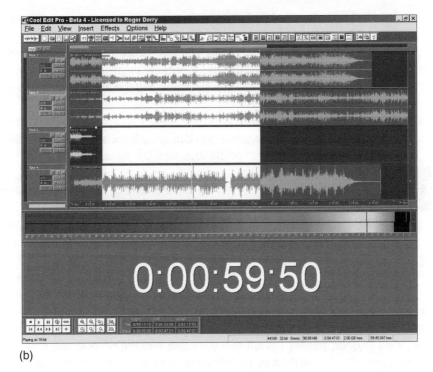

(b)

Figure 17.6 (a) Different layout of multitrack editor; (b) Different layout of wave editor

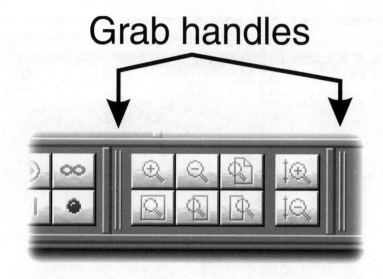

Figure 17.7 Two of *Cool Edit*'s grab handles for moving dockable windows

button. As you move it around the *Cool Edit Pro* interface, two things will happen. If the outline of a dockable window retains its original dimension, releasing the left mouse button will cause the window to 'materialize' as its own standard floating window. However, if you notice the resize bar of another docked window 'light up', this designates a docking location. If this is where you'd like the dragged window to appear, release the left-mouse button and the window will snap into its new location.

To close a docked window, right-click on its handle to see a pop-up menu with a 'Close' option. Check this option and the window will disappear. You can bring back any closed dockable window by checking its name on *Cool Edit Pro*'s View menu.

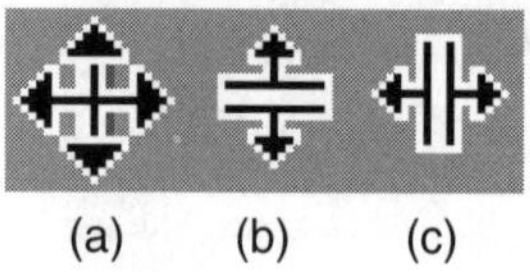

Figure 17.8 (a–c) Mouse cursor appearance when over a grab handle and over vertical and horizontal resizing controls

(a)   (b)   (c)

To move a floating window, left-click on its title bar and, while holding down the left mouse button, drag the window to the desired location. Close a floating window by left-clicking on the 'X' control in its title bar.

To dock a floating window, left-click it and drag it while holding down the left mouse button. As you move it around the *Cool Edit Pro* interface, potential docking locations will appear; you'll see the resize bar of dockable windows 'light up' wherever docking is possible. If this is where you'd like the floating window to dock, release the left mouse button and the window will snap into its new home. If you press the Ctrl key while moving a floating window around, then it will be prevented from docking.

17.8 Device properties

It is here that the input and output of the single wave view editor can be selected to different sound cards if required by selecting the card and ticking the 'Use this device in Edit View' box (Figure 17.9). You can also view what formats your card supports.

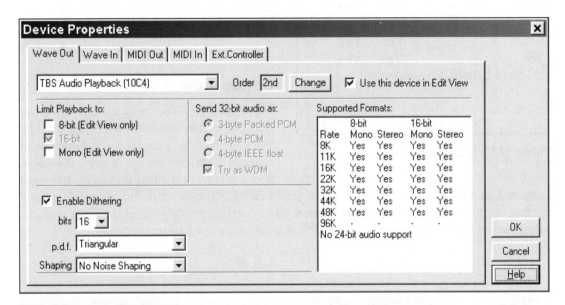

Figure 17.9 Device properties

Professional cards often just support the standard *Pro* bit rates at 16-bit or greater depth (Figure 17.10).

You can also see what order the device appears in *Cool Edit*'s list. You can change this by clicking the change button. This takes you to the same dialog that you get from the menu 'Options/Device order' described in the next section.

There are also options to enable you to handle higher bit-rate files than your sound card is capable of. Optionally you can add dither to 'fill in' the waveform to improve the apparent quality when a 16-bit file is played through an 8-bit card.

The 'Wave In' tab has the option to correct for any DC offset while recording. Quite a lot of sound cards have a DC offset on their output – they shouldn't, but they do – and ticking the 'Adjust to zero DC when recording' box fixes this. Leaving the DC means that you will have constant problems with clicks, especially at the end of files. CDs, for example, will usually go to digital silence in the gaps between tracks, and a click will be caused by the sudden transition from the DC level to zero.

Supported Formats:

Rate	8-bit Mono	8-bit Stereo	16-bit Mono	16-bit Stereo
8K	-	-	-	-
11K	-	-	-	-
16K	-	-	-	-
22K	-	-	-	-
32K	-	-	Yes	Yes
44K	-	-	Yes	Yes
48K	-	-	Yes	Yes
96K	-	-	Yes	Yes

32-bit (4-byte PCM) supported

Figure 17.10 Supported formats

The Multitrack Latency is a fix for the delay within most sound cards when recording. This shows up when you are playing back one track while recording another. You can find that the two tracks are out of synchronization by some milliseconds. This delay is constant for each card. Once you have worked out what it is, largely by trial and error, you can put the correction in and the tracks will line up.

Device order

The order of the devices in *Cool Edit*'s list at first glance may seem trivial (Figure 17.11). The order is used by the track input and output selectors. Provided the list remains the same, then the order is not that important. It is most likely to be useful when you change or add sound cards. You can adjust the allocation so that existing session files do not have to be altered. An important consequence of the order is that it selects which sound card will be the default one when you open a new session file.

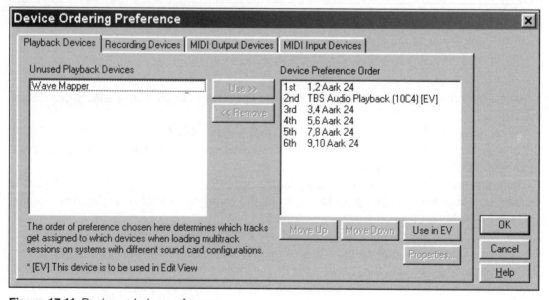

Figure 17.11 Device ordering preference

17.9 Help

This book is not intended to be a *Cool Edit Pro* manual – Syntrillium have already written that. It even comes in printed form when bought as a CD rather than a download. The program does come with substantial help files, which will hold your hand through the various options.

17.10 Icons

Some people love icons, others stare at them uncomprehending. If you like icons, then you can choose which groups of icons for shortcuts will be shown at the top of the screen. At the very least you will probably want to keep the file group, as this includes the icon to toggle between single wave view and multitrack. If you hate icons, then you can remove them altogether. Normally the easiest way to tweak the icons is right-clicking on the toolbar, but if you have no icons selected this is not possible and you need to use the Options/toolbars menu item.

Appendix 1

Time code

All modern audio systems have a time code option. With digital systems, it is effectively built in. At its simplest level it is easy to understand; it stores time in hours, minutes and seconds. As so often, there are several standards.

The most common audio time display that people meet is on the compact disc; this gives minutes and seconds. For professional players, this can be resolved down to fraction of second by counting the data blocks. These are conventionally called *frames*, and there are 75 every second. This is potentially confusing, as CDs were originally mastered from three-quarter-inch U-Matic videotapes where the data were configured to look like an American television picture running at 30 *video* frames per second (fps).

The need accurately to edit videotapes drove the requirement for a standard. VT editing is done by copying from source tapes to the final edited version. With modern microprocessors, tape synchronizers and control gear, this can now be done to frame accuracy (subject to some technical constraints outside the scope of this book).

In 1967, the Society of Motion Picture and Television Engineers (SMPTE) created a standard defining the nature of the recorded signal and the format of the data recorded. Data are separated into 80-bit blocks, each corresponding to a single video frame. The way that the data are recorded (*biphase modulation*) allows the data also to be read from analogue machines when the machines are spooling at medium speed, with the tape against the head, in either direction. With digital systems, the *recording* method is different but the *code* produced stays at the original standards.

The European Broadcasting Union (EBU) adopted the same standard, using the European TV 25 fps frame rate. The core of the format is the actual time code expressed in 24-hour clock mode of HH:MM:SS:FF. It also has eight groups of four *user bits* that the user can decide how to use. The two have been combined in *BS 6865:1987/93: IEC 421:1986/90*.

SMPTE/EBU time code can be recorded as audio on a track of a multitrack tape machine. By convention, the highest numbered track is used; track 4 on a 4-track; track 16 on a 16-track, etc. It is a nasty screeching noise best kept as far away from other audio as possible.

MIDI

Time code can also be sent to a sequencer (via a converter) as MIDI data, allowing the sequencer to track the audio tape. To save too much data overhead, MIDI time code (MTC) is sent only every two frames and even then uses up just over 7 per cent of the available MIDI data capacity. The basic time code is sent as eight separate 2-byte MIDI messages. This also includes a code indicating what frame rate is being used. There are four options: 24 fps (cinema film), 25 fps (video and film for European TV), 30 fps drop-frame (USA/Japan video), or 30 fps non-drop frame (used rarely for non-video applications). Sending other information, such as user bits, is optional. It is important to choose the format that is appropriate for the medium for which you are creating the audio. In the UK, the most likely standard is 25 fps.

Film has 24 separate pictures per second, and these individual pictures are known as frames. Television has adopted a different practice. A European 625-line picture has 25 frames per second, but these frames are divided into two $312\frac{1}{2}$-line fields, which are interlaced together to make the final picture. This doubles the frequency of flicker from 25 Hz to 50 Hz and makes it less visible. Originally, with early technology, TV pictures had to be the same frequency as the local mains to avoid moving 'hum bars'. This was why 25 fps rather than film's 24 fps was chosen. Because USA mains is 60 Hz, their picture standard is 30 frames per second (60 fields per second).

The simple relationship between bars, tempo and SMPTE time as shown by sequencers like *Cubase* is only valid for 120 beats per minute 4/4 time. MIDI time-code generators need to be programmed with the music tempo and time signature used by the sequencer, so that they can operate (in a gearbox fashion) so that the sequencer runs at the proper tempo.

Drop-frame time code

This used in the USA to deal with a fudge that was made when they introduced colour TV. This involves dropping frames so that time code agrees with real time. While American B&W TV ran at 30 fps, the colour frame rate is nearer 29.97 fps. This means that a programme timed at an hour using 30 fps time code, will actually run 3.6 seconds longer on colour TV. This may not seem much, but it worries broadcasters. The SMPTE decided to standardize a way that the 108 'extra' frames could be 'dropped' every hour. It's similar to the way that the calendar has leap years to keep it in synch with the orbit of the Earth.

The basic rule is that whenever the time code ends a minute, it drops the first two frames on the next minute. For example, 12:25:59:29 is followed by 12:26:00:02 – frames 00 and 01 are dropped. Because this would drop 120 frames rather than 108 frames per hour, there is a further tweak to the rule, so that every tenth minute does not drop any frames – thus 12:29:59:29 is followed by 12:30:00:00. This makes drop-frame time code accurate to 75/1000th of a second in 24 hours, which is good enough for most people.

Other systems

Modern digital systems like CD, DAT and Minidisc all come with their own internal time codes. Their internal digital audio data is divided into blocks also called frames. Their connection to 'outside world' time code systems will translate to the SMPTE standard. DAT, for example, stores any incoming SMPTE time code in an internal format that it can then output in any frame rate regardless of the original.

Film has traditionally used a totally different system for dubbing, known as 35-mm feet. This increments faster than every second. As you would expect, this corresponds to the number of feet of 35 mm film, but the measure is also used when the film is actually 16 mm!

What time?

Time of day recording can be useful when recording an event. Multiple cameras can stop and start and everything be pieced together later. Production assistants can log events using an ordinary accurate clock or watch.

In practice a variation on duration coding is used, but it is always unwise to start your recording at zero. There is no negative time, so this would give no space for a pre-roll to allow times for separate equipment to synchronize. Very often the 'hours' are set to indicate which film roll or videotape is being used – roll 1 will be given a 01:00:00:00 start time, roll 2 gets 02:00:00:00, etc.

The 80 bits of each SMPTE/EBU frame block are allocated as follows:

Bits 0–3	Frame units
Bits 4–7	User bits
Bits 8–9	Frame 10s
Bit 10	Drop frame flag/NU
Bit 11	Colour frame flag
Bits 12–15	User bits
Bits 16–19	Second units
Bits 20–23	User bits
Bits 24–26	Seconds tens
Bit 27	Group flag 2/parity
Bits 32–35	Minutes units
Bits 36–39	User bits
Bits 40–42	Minutes tens
Bit 43	Group flag 2/0
Bits 44–47	User bits
Bits 48–51	Hours units
Bits 52–55	User bits
Bits 56–57	Hours tens
Bit 58	Group flag 1
Bit 59	Parity/group flag 2
Bits 60–63	User bits
Bits 64–79	Sync. word (0011111111111101)

Appendix 2

Clicks and clocks

When you copy digitally to your computer, are you plagued with clicks and plops? Is there a regular beating sound where there should be digital silence? If so, you are probably suffering from unsynchronized word clocks. Arcane though it sounds, the problem is simply solved.

Copying digitally from one device to another is, in many ways, easier than using an analogue link. There is no level adjustment to be made and quality is assured, provided the link is made by good quality cable.

Yet there is an extra factor that it is vital to get right, and this is what is known as word clock synchronization. You will be using 32 000, 44 100 or 48 000 samples per second sampling. These figures are also known as the word clock rate – 48 000 samples per second also means 48 000 digital words per second. For a 16-bit stereo signal each word will consist of two 16-bit samples (left and right) packaged together with additional 'housekeeping' data.

The word clock synchronizes the whole system; it can be thought of as the conductor of an orchestra. When you are copying from digital recorder to digital recorder, there are no complications. It is only when you have a device handling several digital signals at the same time – like a digital mixer or a computer sound card – that problems arise.

Returning to the conductor of an orchestra analogy, there are some pieces of music – Mahler symphonies, etc. – that feature offstage bands that cannot see the conductor in the hall. Unless something is organized, they will not be able to play in time with the hall orchestra. A common modern solution is to have a closed circuit TV connection, so that a second conductor can synchronize his beat to the image of the conductor in the hall.

So it is with digital systems. If your card can handle other inputs at the same time as the digital one, then you need to be concerned with word clocks. The 'conductor's beat', *aka* the word clock, is sent as part of the digital signal – so we have our 'closed circuit TV'. Normally the sound card will generate its own 'beat'. This you set when you choose the sampling rate. While this is set to match the rate coming in from the external playback, there is no synchronization between the two. Even with crystal control the chances of both clocks being precisely the same frequency are remote, so every few seconds they drift one word apart. The error resulting from this is what causes the clicks or plops. The answer is to tell the sound card to 'look' at the word clock incoming from the DAT (or Minidisc) machine feeding it. The sound card management software should have this as an option, alongside the sampling rate selection.

Figure A2.1 shows dropdown menus for a multitrack card that can handle either optical or electrical digital inputs (the selection between the two is made in a separate menu).

If you select the S/PDIF or TOSLINK option, then the sound card is controlled entirely from the external bit stream; it will follow the sampling rate as part of the synchronization.

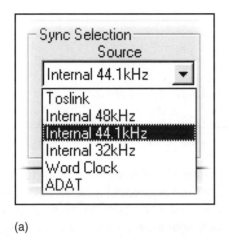
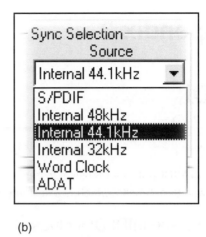

(a)   (b)

Figure A2.1 Dropdown menus for a multitrack card: (a) Toslink; (b) S/PDIF

Beware: you have to switch synchronization back to internal once you stop using the external machine. As soon as it is disconnected you will have no clock, and the sound card may either not produce any sound, or produce sound at the wrong rate and pitch.

Another trap is that an external machine may default to a different clock when it is not playing back. This means that, having very successfully copied a 44.1-kHz recording, the clock may go to 48 kHz when you take the DAT recording out. If you have not reset your sound card back to an internal clock setting, then the files played back through your editor may well play at the wrong speed and pitch – in this case, faster and higher.

Word clock in/out

The illustrated sound card can also synchronize to an incoming ADAT 8-track feed. It has another option, which is specifically labelled word clock. This is for a fully professional installation where a separate high quality word clock generator is used. This separately synchronizes a number of separate digital devices – essential when dealing with a digital mixer.

Equally, when you have more than one digital sound card you have to synchronize them by connecting the word clock output of the card you have designated as the master to the word clock in of the second card. The second card will always synchronize to the master provided it is set to use the word clock input, including following the sampling rate settings. If a third or fourth card is added, then the word clock can be 'daisy-chained' from output to input of the next. Word clock connections are usually made with professional video-style BNC connectors. They keep the cards in synch even when there are no audio data present.

Appendix 3

MIDI

The term MIDI stands for Musical Instrument Digital Interface. Do not confuse it with the term midi as applied to hi-fi units, which is a description of their size (midi as opposed to biggi). MIDI is what is known as a serial connection. The data are sent as a series of long and short pulses in groups of eight. Additionally there is a pulse on either side known as 'start' and 'stop' bits. Just over 30 000 of these pulses can be sent every second – 3000 'groups' known as 'bytes' (31.25 kbaud in tech-speak). Figure A3.1 shows the difference between audio and MIDI connections.

Audio and MIDI DIN plugs

Stereo Audio   Mono Audio   MIDI

MIDI Sockets

IN   OUT   THRU

Figure A3.1 Top, the difference between audio and MIDI connections (white pin indicates not used); Bottom, standard MIDI sockets

The connections are via DIN plugs, which although physically identical to audio DIN plugs are differently wired (Figure A3.1, top).

A typical synthesizer will have three sockets as shown. The 'IN' will drive the synthesizer (sound making) circuits. The 'OUT' will carry the output of the keyboard. The MIDI system allows up to 16 instruments to be connected together in a 'daisy chain' (Figure A3.1, bottom). The 'THRU' socket (American spelling of 'through') relays the data arriving at the IN socket, unchanged, onto the next instrument. In the process the electrical signal is 'cleaned up' and isolated so that a fault on one machine will not necessarily adversely affect the rest of the instruments in the chain (Figure A3.2).

In general, to send each command to an instrument takes 3 bytes of data. There is the possibility that an individual instrument may get sent too much data for it to cope with. Because of the 'live' nature of performance, it is vital that it can recover from this very quickly and does not 'go out of synchronism' with the data. This is achieved by uniquely marking the bytes of data. The first byte of a group of three is the 'command' byte, and has three sections. The bit corresponding to the highest value represented is always ON, and this means that only numbers from 128 to 255 are sent. The 4 bits that correspond to the lowest values (0–15) are allocated to indicate which machine the command is for. Because people are not used to the concept of 'machine 0' the channels numbers are named 1–16, although the actual numbers sent are 0–15. The remaining 3 bits are used to define the command.

Engineers notate the numbers represented by these 3 bits combined with the eighth bit, which is always set as 8, 9, A, B, C, D, E, F (hexadecimal). The most important commands are 'note on' = 8, and 'note off' = 9.

The other two bytes are number values from 0 to 127. The eighth bit of a data byte is always 0 to indicate that the byte is data, not a command. What these data represent will depend on the command, but a typical example is 'note on'.

In Figure A3.3, the first data byte is a number that represents the note to be played on the standard western music scale where there are 12 notes to an octave (including the black keys). Thus two note numbers that are different by 12 are an octave apart. Other commands can modify the pitch for special effects. The second data byte represents the velocity – how 'hard' the key is pressed.

In practice most synthesizers are polyphonic; they can sound more than one note at a time and can respond to more than one MIDI channel at a time. They may also contain drum machines etc. with their own MIDI channel number.

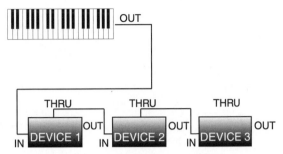

Figure A3.2 'Daisy-chaining' MIDI devices

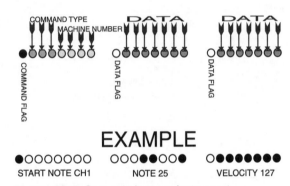

Figure A3.3 Commands to an instrument

Synthesizers usually send out a data 'heartbeat' several times a second. This can distract computers and sound cards when being switched on, causing lockups or crashes. Unless you are sure that your machine is free from this problem, it makes sense to switch off or disconnect your synthesizer when you switch on your computer at the beginning of a session.

Glossary

AB stereo   Often used to distinguish MS stereo from the convention signal using left and right signals. In the context of microphones it often implies the use of spaced omnidirectional mics rather than a coincident pair.

ADAT   A digital multitrack recording system that gives eight tracks on an S-VHS videocassette.

ADPCM (Adaptive Differential Pulse Code Modulation)   Conventional Pulse Code Modulation stores the values of a waveform as a series of absolute values. Differential PCM does not do this but instead sends the data as a series of numbers indicating the difference between successive samples.

ADSL (Asymmetric Digital Subscriber Line)   Sometimes known as a 'cable modem', this is telephone data service where higher data speeds are available. The connection is permanently open. The 'Asymmetrical' refers to the fact that the data rate is slower for uploads compared with downloads. The actual data bandwidth that is available is shared between a number of subscribers (contention ratio), and will varying depending on how many are sending or receiving data at any moment.

AES (Audio Engineering Society)   Among other things, the AES lays down technical standards. In the context of this book they are best known for a professional standard for conveying digital audio from machine to machine, which has also been adopted by the EBU.

AES/EBU   A professional digital audio standard for transferring digital audio between machines. This is balanced and uses XLR connectors. The data format is similar but not identical to S/PDIF, which can see an incoming AES/EBU format signal as copy prohibited.

AIFF   Apple AIFF (.AIF, .SND) is Apple's standard wave file format, and is a good choice for PC/Mac cross-platform compatibility.

ALC (Automatic Level Control)   See AVC.

Aliasing   Spurious extra frequencies generated as a result of the original audio beating with a frequency generated within the audio processing system (usually the sampling frequency in a digital system). Filters are used on the input to prevent this, but these filters themselves can produce degradation of the signal unless very well designed.

Analogue audio   Until digital techniques came along audio was conveyed and recorded by using a property that changes 'analogously' to the sound pressure. This property might be electrical voltage, magnetization, or how a groove wiggled. Digital audio replaces this with a series of numbers.

AV (Audio Visual)   AV standard hard drives are able to cope with long runs of data (such as a long continuous audio recording) without stopping to recalibrate themselves for temperature variations.

AVC (Automatic Volume Control) Often found as an option on portable recorders, this automatically adjusts the recording level from second to second. This can cause trouble when editing, as the background noise will be going up and down. However, modern AVCs work surprisingly well. The background matching becomes a trivial problem when editing with a PC digital audio editor.

Balanced Normal domestic audio connections are unbalanced; a single wire carries the audio, which is surrounded by a screening braid connected to earth as the return circuit. These circuits are prone to pick-up of unwanted signals as well as high frequency loss when used beyond about 5 metres. Balanced circuits use two wires to carry the audio (still within a screening braid). The audio in the wires is going in opposite directions – as one wire goes positive the other goes negative. Interfering signals induce in the same direction on both wires. The input circuit is designed only to be sensitive to the difference in voltage between the two wires and therefore ignores the induced interference.

Barrier mic Barrier mics are designed to be placed on large, flat surfaces rather than suspended in free air. These are often referred to as PZMs (Pressure Zone Microphones) after a commercial version.

bel See Decibel.

Binaural Current practice is to use this term to mean two channel audio balances intended to be heard on headphones.

BIT (BInary digiT) Digital/PCM systems use pulses that indicate either an 'ON' or 'OFF' state. Each individual piece of data is known as a bit.

BNC A professional video/digital audio connector with a locking collar.

Byte A group of 8 bits (allegedly a contraction of 'By Eight'). This is the standard measure of the capacity of digital systems.

Capacitor An electrical component that can store electrical charge, formerly known as a condenser. They consist of two parallel 'plates' separated by an insulator. The plates are so close together that when they are charged the positive charge on one plate is attracted to the negative charge on the other. The closer they are together, the greater the attraction. This increases the amount of charge that the device can store. A practical capacitor's plates are in fact metal foil sheets separated by a sheet of thin insulator rolled into a cylinder, rather like a Swiss roll. This reduces their size and gives them a cylindrical appearance. A specially constructed capacitor forms the basis of electrostatic microphones.

Cardioid (heart-shaped) The most common of microphone directivity shapes. (See page 30.)

CD (Compact Disc) When used without qualification this is taken to mean a standard audio CD. Subsequently adopted for data use in computers, this has led to many variants (see below).

CDR (Compact Disc Recordable) A recordable CD that cannot be erased.

CD-ROM (Compact Disc Read Only Memory) A CD containing data rather than audio.

CD-RW (Compact Disc Read-Write) A recordable CD that can be erased. While these discs can be recorded in audio format, most domestic CD players cannot play them.

Chequerboarding A technique for mixing sections of audio or video. (See page 90.)

Cinch plug Another name for phono plug.

Clock All digital systems have a reference clock, which acts rather like the conductor of an orchestra to keep everything in sync. (See Appendix 2.)

Clone   In the digital audio context this is used to mean making an exact sample for sample copy. This is not possible with systems like Minidisc, which use lossy compression systems.

Coax, **Coaxial plug**   A generic term for a connection that uses a cable where one or more conductors are surrounded by a wire braid that helps screen out interference. In the UK, this term is most often used for the plug used for television aerials. In the audio context, some companies use this term for a phono plug.

Coincident pair   A stereo microphone technique using directional mics placed as close together as possible.

Compact cassette   The proper name for the ordinary analogue audio cassette; undoubtedly the most successful audio recording medium ever invented.

Compression: Audio   Compressor limiters are the most used effects devices in the studio. They can be thought of as 'electronic faders' that are controlled by the level of the audio at their input. (See page 112.)

Compression: Data   There are two types of data compression; non-lossy and lossy. *Non-lossy* compression is the first (and traditional) form of compression, and reduces the data to be stored on a disk or sent via a modem. The best-known format is the Zip format. A Zip file can be uncompressed to recreate the original data without any change or error. *Lossy* compression is when graphics and audio are often compressed using formats that approximate the data using assumptions about how we see and hear. (See page 181.)

Condenser microphone   Condenser is an old term for capacitor.

Cool Edit Pro   A commercial audio editing package for Windows, combining both linear and non-linear editing.

DAC   Digital to analogue converter

DAO (Disc At Once)   A technique of burning a CD in one go. This has a number of technical advantages, notably giving the ability to control inter-track gaps or even recording audio into those gaps.

DAT (Digital Audio Tape)   Originally a generic term for 'Digital Audio Tape Recorder'. This is now used specifically for a format developed by Sony, and supported by scores of other manufacturers, that has become popular amongst professionals and semiprofessionals alike for mastering digital audio.

Data   Plural of 'datum'. From Latin 'things given'. The word is often erroneously used as a singular (correct use is 'The data are' rather than 'The data is').

DAW (Digital Audio Workstation)   A dedicated computer audio editor with specialized controls and software.

dB   Decibel

dBA   Decibels are used widely in the field of acoustics, audio and video. Subtly different scales are used, and the different types of decibel are indicated by a suffix. The dBA is used in the field of acoustic measurement.

dbx   A commercial company that is best known for developing a popular analogue noise-reduction system.

DC Offset   Poor analogue to digital converters can have a DC offset that 'pushes' the audio away from being centred on zero volts towards either the positive or negative. This is a major cause of clicks on edits. (See section 6.3, Figure 6.4.)

Decibel One-tenth of a bel. This is the normal way of measuring audio. It is a logarithmic system using a standard reference level, and values are expressed as a ratio of a standard level. The bel itself is too large a unit to be convenient for audio. The Richter units used for measuring earthquakes are identical to bels, but with a rather louder reference level! (See page 5.)

Delta modulation A technique where the difference between samples is sent instead of the absolute values usually sent by PCM.

Digital audio Sound pressure level variations are represented by a stream of numbers corresponding to pulses.

Digital audio workstation A dedicated computer audio editor with specialized controls and software.

Digital Versatile Disc (DVD) This is the likely successor to the compact disc. It uses similar technology but takes advantage of the technical developments since the CD was introduced. Recordings are made at a much higher density – eight times greater than CD. Additionally, the DVD is double-sided and also each side can be made up of two layers. This gives massive data capacity, enough for full-length films with 5.1 channel audio. The '.1' is a (not very good) engineering joke – there are in fact six channels of audio, but the sixth channel is a low bandwidth one used for low frequency effects.

DIN (Deutsche Industrie-Norm) German industrial standard. This includes the audio/MIDI/ computer DIN plug, which is a standard-sized case and connector containing a number of pins. The most common type met in the audio context is the five-pin 180°.

Disc, Disk A convention has grown up where disc-based media using a magnetic medium are spelt as 'disk'. Optical recordings like CD and Minidisc, as well as gramophone records, remain spelt as 'disc'.

Dither A low-level signal, usually random noise, which is added to the analogue signal before conversion to digital. Its effect is to reduce the distortion caused by quantization.

Dolby Dr Ray Milton Dolby, possibly the most influential individual in audio. His company, Dolby Laboratories, began by making audio noise reduction systems. The first, a professional system, became known as Dolby A-type. A simplified system, called Dolby B-type, revolutionized the compact cassette medium for consumers. Dolby C-type gives about 20 dB noise reduction compared with the B-type system's 10 dB, albeit at the cost of poorer compatibility when played on machines without decoding. Dolby SR (Spectral Recording) is an enhanced professional system that can give better than 16-bit digital performance from analogue tape. Dolby S is a powerful compact cassette system based on a simplified version of SR. As other companies have to obtain a Dolby trademark, the company has become the effective setter of standards for cassette machines, as minimum audio performance is set by Dolby Laboratories to allow their systems to be used.

Dolby Digital Started in the cinema industry, this is the digital multi-channel sound format most widely used on DVD and with digital television.

Dolby Stereo Dolby Laboratories devised a system for putting stereo onto optical release prints in the cinema. This was combined with a phase coding system that allows surround information to be heard. The Left total and Right total channels can be decoded to give Left, Centre, Right and Rear loudspeaker information. In the early 1990s the term 'Stereo' was dropped, and the analogue film format is now simply referred to as 'Dolby'.

Dolby Surround The domestic version of Dolby stereo, as found on video cassettes, CDs, TV broadcasts and video games.

Drop in Switching from play to record while running to make an electronic edit.

Drop out (1) A momentary loss of sensitivity in an analogue recording medium. Digital systems can correct or conceal errors resulting from drop outs in the medium. However, if they fail you will hear a mute instead. (2) Switching from record to playback to end a drop in.

DTMF (Dual Tone Multi-Frequency) These are the tones that modern phones make when dialling. They are used to code the digits 0–9 as well as the special system codes of '*' and ''. Four extra codes are also available, known as 'A', 'B', 'C' and 'D'. The sixteen possible combinations are achieved by sending two frequencies (hence 'dual tone') out of a possible selection of a total of eight.

DVD Digital Versatile Disc

Dynamic microphone An alternative word for a moving-coil microphone.

Dynamic range In audio systems, the dynamic range available is determined by the number of bits used to measure each sample. In theory, for every bit extra another 6 dB of signal-to-noise ratio is gained.

EBU (European Broadcasting Union) A trade association for European broadcasters, which also sets technical standards.

Echo Although this is often used interchangeably with the term reverberation (or 'reverb'), they are technically different. Echo is where you can distinguish individual reflections (Echo . . . Echo . . . Echo . . . Echo) while reverberation is where there are so many reflections that they merge into one continuous sound. (See page 116.)

Edit Decision List (EDL) Used in a non-linear editing process, where the audio files are not altered. Instead a list of instructions (the Edit Decision List) as to what section to play when, at what level, etc. is created, which causes the hard disk to skip around and produce apparently continuous audio.

EDL Edit Decision List

EIDE (Enhanced Integrated Drive Electronics) The most widely used (and hence cheapest) form of hard drive. Most PC motherboards are equipped to handle four drives. The usual alternative is SCSI.

Electret A form of capacitor that remains charged permanently. This means that when constructed as a microphone, it does not need a polarizing voltage. Typically, a simple 1.5-V AA battery is used to power the built-in amplifier.

Electromagnetic Describes devices where magnetism is used to create electricity or electricity is used to create magnetism. Within audio, a moving-coil (dynamic) mic is used to create electricity (the audio signal) by the diaphragm pushing and pulling a coil of wire between the poles of a magnet.

Electrostatic Electrostatic microphones use a diaphragm that is one of the plates of a capacitor held charged by a polarizing voltage. (See page 34.)

EQ Pronounced 'Eee' 'Cue'. A widely used abbreviation for 'equalizer', a term for devices like tone controls that modify the frequency response of an audio system. Such devices were originally used by engineers actually to equalize or correct audio deficiencies in land lines and recording systems. They were then borrowed by studio operators to improve their recording mix. Nowadays equalizers are designed specifically for studio use.

Error concealment   A technique where errors can be detected but not corrected. Instead they are concealed, often by replacing the sample with an average of the samples either side. This is called interpolation (see below).

Error correction   As digital audio is a series of numbers, extra numbers can be added having been generated by various mathematical means. At the receiving end, these numbers can be generated again from the incoming data. If they are different from the extra numbers sent, then error has been detected. With suitable maths, the errors can often be corrected. Because the binary nature of the signal represents only '0' or '1', it is clear that if the incoming value is established as being wrong, then the correct value must be the only other value. *Interpolation*: a technique where when an error is detected a value between the preceding and following value is substituted, whereas full error correction actually reconstitutes the data. *Interleaving*: errors can be made easier to correct by interleaving the data so that they are physically spread out on the medium so that a single drop out in the medium does not produce a single burst of errors.

fff   'f' is a music term for loud (Italian: *forte*); the degree of loudness is indicated by the number of 'f's, thus fff is very loud.

FFT (Fast Fourier Transform)   A mathematical way of defining a filter.

Figure of eight   Term used to describe a microphone that is sensitive at the front and back but dead at the sides.

Firewire   A standard way of connecting apparatus to a computer that is supported by both PCs and Apple Mac computers. The devices are 'daisy-chained' together and can be connected or disconnected without having to reboot the computer. This is faster, although presently less common, than USB.

Flanging   Phasing with continuously varying delay.

Flutter   Rapid variation of pitch, often caused by a dirty or damaged capstan pulley on an analogue tape machine. A digital recording will be free from this fault.

Flutter echo   Term used by those who tend to use the terms 'echo' and 'reverb' interchangeably, to show that they are really talking about echo.

FM   (Frequency Modulation)   A way of sending data or audio using a carrier frequency that is varied in pitch. This is used by FM radio, VHS video cassette, hi-fi sound, many hard disks, etc.

FX   A widely used abbreviation for 'effects'.

Gain   Another word for amplification.

Giga   1000 million – hence 1 GigaHertz = 1 000 000 000 Hertz = 1 GHz.

Glitch   A discontinuity in sound, due to data errors.

Gun Mic   A generic term for very directional mics that use a long tube and phase cancellation techniques. Also known as shotgun or rifle mics. Because they are very sensitive to wind noise they are usually concealed in long, furry, sausage-like windshields.

Hexadecimal   A numbering system based on the number 16 instead of 10. The characters 0–9 and A–F are used – 0, 1, 2, 3, 4, 5, 6, 7, 8, 9, A, B, C, D, E, F, 10, 11, 12 etc. It is a convenient notation for binary numbers as used by computers and for MIDI. FD is easier for a person to distinguish from FB than 11111101 from 11111011.

Hypercardioid   A microphone that has a slightly narrower front pick-up compared with a cardioid. The penalty is that there is a reduced sensitivity lobe at the back, giving a dead angle a few degrees to the side of this.

IDE (Integrated Drive Electronics)   A way of connecting hard drives to personal computers. It is very cheap and has become very popular, and hence even cheaper. Although the technology has much improved it is regarded by many as inferior to SCSI, as it uses the computer's processor and slows down programs that are being run at the time. It has been said that IDE works in a way analogous to arriving at a shop counter, asking for an item and saying 'I'll wait', whereas SCSI allows you to go away and do something else because it will 'deliver'.

Image   The perceived location of a single source within the sound stage. The image may be narrow (panned mono) or wide (string section). It may be precise or blurred.

Interleaving   Error correction technique (see **Error correction**   above).

Interpolation   Error concealment technique (see **Error correction**   above).

ips   Inches per second.

ISDN (Integrated Services Digital Network)   Sometimes ironically referred to as the 'It Sometimes Doesn't Network', this gives you a direct digital connection to the phone network. There are various audio coders that allow good quality down what is effectively a telephone line. Different organizations have standardized on different systems, and some manufacturers' gear is not fully compatible.

Kilo (K)   The word kilo when used in the context of digital systems usually means 1024, *not* 1000. This represents the maximum value of a 10-bit word, i.e. 2 to the power of 10. This is not cussedness, but is used because it represents a very convenient unit. So the term 64K will mean 65 536. Some publications use the abbreviation 'K' to mean 1024 and 'k' to mean 1000.

Limiting   Compression of greater than $10:1$.

Line in   An input to be fed by amplified audio rather than a microphone.

Line level   Domestic outputs tend to be $-8\,dB$, with professional outputs being $+4\,dB$. $0\,dB$ is 0.775 volts. This strange value dates from telephone technology, where 0.775 volts gave 1 milliwatt into a 600-ohm circuit.

Line out   Amplified output of a device.

Linear editor   An editor where the audio files themselves are altered by the editing process.

Lossy/non-lossy   See **Compression: Data**   above.

LP (Long Play)   Usually used to refer to 12-inch gramophone records. However, video cassettes, DAT and Minidiscs have a long play mode.

Mastering   The general term for transferring studio material to a final stereo 'master', which will be used to generate the copies sold to the public.

Mega   1 million, hence 1 megahertz = 1 000 000 Hertz = 1 MHz (note upper case 'M'; lower case 'm' means milli = 1/1000th when used as a prefix).

Mic   Preferred British abbreviation of the word microphone.

Micro   One millionth, hence 1 microsecond (1 μs). Used colloquially to mean small computer.

MIDI (Musical Instrument Digital Interface)   A standard system for communicating performance information between synthesizers and computers.

Milli   One thousandth, hence 1 millimetre = 1 thousandth of a metre (1 mm).

Monaural   Sometimes used to mean monophonic, but really means listening with one ear!

Monitoring speaker   Monitoring speakers are designed to be analytical and reveal blemishes so that they can be corrected. As a generalization, hi-fi loudspeakers make the best of what is available.

Mono   Contraction of monophonic; also contraction of monochrome (i.e. black and white pictures).

Monophonic   Usually contracted to 'mono', this is conventionally derived from the stereo signal by a simple mix of left and right channels. Beware of some portable tape machines that have a switch labelled mono. This often means that only one input is fed to both legs of the recording, *not* that the two inputs are mixed and fed to both legs. It is very easy to end up with an interview tape that only has questions on it if a reporter is not aware of this.

Moving coil   The most common type of microphone and loudspeaker. A coil of wire attached to a diaphragm is suspended in a magnetic field. If the coil is moved by pressure on the diaphragm, a small voltage is generated. If a current is passed through the coil, the diaphragm will be moved.

MPEG (Motion Picture Experts Group)   MPEG is a series of lossy compression systems for audio and video.

MS (Middle-Side)   A stereo technique where instead of the left and right information being used for the two audio channels, the middle and side are used. The middle signal corresponds to the mono signal. The side signal corresponds to the amount of 'stereo-ness'. It is zero for centre signals and at a maximum for sounds coming from the extreme left or right.

Multiplex   Generally any method of carrying several signals (e.g. stereo left and right) on a common circuit. Digital radio and television are broadcast using multiplexes that carry a number of channels; how many depends on the quality required.

Nano   1 thousand millionth, hence 1 nanosecond (1 ns).

NICAM (Near Instantaneously Companded Audio Multiplex)   Digital system used in the UK to add stereo sound to television broadcasts.

Noise reduction   Analogue recordings are often made using noise-reduction systems like Dolby and dbx. They boost the signal on record and apply a correction to this on playback. In the process, noise and hiss are correspondingly reduced. In the digital editor context, noise reduction usually refers to various software solutions that remove clicks, hiss or noise from an existing recording – usually a transfer from an analogue original.

Non-linear editor   An editor that does not alter the original audio files. Instead it uses some form of edit decision list to instruct the computer to jump around the hard disk, reproducing and mixing audio as required.

NTSC (National Television System Committee)   American colour television system.

Nyquist limit   This is named after Harry Nyquist, the Bell Telephone Laboratories' theoretician who first enunciated the principle that you have to sample at a frequency at least twice that of the highest you intend to transmit.

Omnidirectional   Omnidirectional microphones are equally sensitive to audio from any direction.

Out-of-phase   If loudspeakers are out-of-phase, this means that their diaphragms move in opposite directions instead of moving in and out together. This has the effect of blurring the sound image and reducing the bass response of a system.

PAL (Phase Alternate Line)   Colour television system used by most countries in Europe. This is often used to imply a 25-fps frame rate, although 30-fps versions do exist. Many video cassette machines will output a PAL signal at 30 fps when playing an NTSC tape.

PC   Originally a generic term for 'personal computer', this has been hijacked to mean a computer based on the original IBM design. While sometimes used to imply using a Microsoft operating system such as Windows (as opposed to Apple Mac), they can also be used for other operating systems such as BeOS, Unix, Linux, etc.

PCM (Pulse Code Modulation)   The technique of sending numbers as a series of pulses.

Phantom volts   Electrostatic (capacitor or condenser) mics need to be powered to work. Balanced studio mics are often powered by adding the DC voltage (usually 48 V) to the audio wires. Circuitry at each end separates the audio. Although the circuit behaves as if there is an extra wire for the power, it has no physical existence – hence the term phantom.

Phase   A measure of the relative delay between two waveforms at the single cycle level. The positive-going zero crossing is described as 0° and the negative-going zero crossing as 180°. Where the difference between the two waveforms is exactly reverse – positive in one is matched by negative in the other – they are said to be 180° out of phase, or just out of phase.

Phasing   This is caused by selective cancellation of some frequencies either using a comb filter or, more often, by mixing two nominally identical signals with a short delay between them. In the late 1960s this was a much-loved 'psychedelic' sound. If the delay is continuously varied, this is called flanging.

Phono plug   The RCA phono plug was originally designed to connect phonograph (gramophone) turntables to amplifiers. This has become a universal way of connecting unbalanced audio (and often video), largely because of the cheapness of the connector.

POTS (Plain Old Telephone System)   A general term for ordinary telephone lines, which have limited capabilities compared with specialized systems using newer technology such as ISDN.

ppp   'p' is a music term for quiet (Italian: *piano*); the degree of quietness is indicated by the number of 'p's, thus ppp is very quiet.

Pre-emphasis   A technique of boosting high frequencies on record or transmission; they are then restored on playback or reception. In the process, any added hiss has top cut applied to it.

Program, programme   In this book 'program' is used in the context of computers. The British spelling is used in the context of radio and TV programmes.

PZM (Pressure Zone Mic)   See Barrier mics.

Quantizing   (1) The process of turning an analogue signal (like audio) into numbers. (2) In MIDI sequencers, the automatic moving of notes onto the beat.

Quantizing interval   The difference in voltage between quantizing levels.

Quantizing levels   The number of possible values into which an analogue signal may be divided or quantized.

RCA plug   The RCA company originally devised a plug for connecting phonographs (gramophone turntables) to amplifiers. These plugs became known as phonograph plugs, abbreviated to phono plugs.

Reverb   Contraction of 'reverberation'.

Reverberation   The sound made by reflections from a room, or an electrical simulation of this. It differs from echo in that there are so many reflections that individual reflections are not discernible. Reverberation time is measured as the time taken for the reflection level to decrease by 60 dB (RT60).

RIAA (Radio Industries Association of America)   A trade association that also sets standards, of which the most often met is the RIAA standard for equalization of long playing gramophone records. When the disc is cut, the high frequencies are boosted and the bass frequencies cut. This is corrected by reversing this on playback.

Ribbon mic   A microphone using a corrugated aluminium ribbon as a diaphragm. This is placed within the field of a powerful magnet. Once the standard microphone used by the BBC and other broadcasters, it is capable of extremely good quality, especially on strings. It is advisable to take off your wristwatch before handling one, as the strong magnetic field may stop the watch.

S/PDIF (Sony/Philips Digital Interface)   Standardized digital audio connection format using an unbalanced audio connection and phono plugs.

SCMS (Serial Copy Management System)   A specification for copy protection flags contained within domestic digital audio connections. While the system allows for no copy inhibition or full copy protection, its default is usually to allow only one digital copy; a SCMS-equipped digital recorder will copy a digital recording set to allow one copy, but the copy it makes will be set to no copying allowed. In practice the system is merely an annoyance for serious recordists working with original material. It can easily be defeated by analogue copying, copying via a computer, or copying via an interface box that allows the resetting of the SCMS flags.

Scrub editing   A term now used to describe the traditional way of finding edits on reel-to-reel tape, which involved 'scrubbing' the tape back and forth. Many users familiar with tape editing like to have this facility on a digital editor, although this usually soon turns out to be a security blanket.

SCSI (Small Computer Systems Interface)   Usually pronounced 'scuzzy', this is a way of connecting personal computers. Originally used on Apple Macintosh computers, it is regarded as having many technical advantages over its rival IDE, which is universally used by PCs. However, there is no reason why a PC cannot use SCSI (some PC motherboards have it built in, and others can have a SCSI card fitted). Many people regard the extra expense worth it for the greater reliability it gives to audio recording. Both systems can be used within the same machine. The SCSI interface is also used for other devices like scanners. Seven devices plus the controller can be accommodated by a simple SCSI system, with fifteen devices available on more sophisticated systems.

SECAM (System En Couleur Avec Memoire)   French Colour Television system. SECAM VHS tapes will play in monochrome on a PAL video cassette machine.

Serial   A term used when data is sent down one wire, one bit at a time.

SMPTE (Society of Motion Picture and Television Engineers)   In conversation often referred to as 'simty', this body sets standards.

SMPTE/EBU   The combined American and European time code standard.

Stereo, stereophony   Literally 'solid sound', from the Greek. While there is argument over a precise definition, it seems generally to mean using two channels to give a directional effect. Good stereo will do more than this, giving a sense of depth as well as direction.

Table of contents   Compact disc and Minidisc use a table of contents to tell them where each track and index begins and ends. If this is lost, then the whole disc becomes unplayable.

TAO (Track At Once)   Where a CD is burned one track at a time with one wave file per track. The laser is turned off between tracks.

Time code   A code contained within a recording that identifies each part of the tape uniquely in terms of time.

Tinnitus   A distressing condition caused by hearing damage, where noises are generated within the ear and are often perceived by the victim to be at a very loud level.

TOSLINK (TOShiba LINK)   Optical connector used by many domestic digital audio devices.

Transient   A short-lived sound, typically the leading and trailing edges of a note.

USB (Universal Serial Bus)   A standard way of connecting apparatus to a computer, which is supported by both PCs and Apple Mac computers. The devices are 'daisy-chained' together, and can be connected or disconnected without having to reboot the computer. The presently less common Firewire system is operationally similar but faster.

Variable pattern mic   This contains two mic capsules that can be combined in different ways to give a range of directivities. (See page 32.)

Varispeed   A control for varying the speed of an analogue recorder, usually on playback. This not only changes the rate but also the pitch. Digital editors usually have software that will allow changes to rate or pitch independently of each other. Small changes of less than 10 per cent are usually very successful, but larger changes can suffer from glitches and artefacts. As a result you are often offered a choice of software methods (algorithms) that have different strengths and weaknesses.

VHS (Victor Home System)   Presently the most popular home video tape format. In its NTSC version, this has long play and extended play modes in addition to standard play. PAL systems only have a long play mode. The audio is recorded in two forms. There is a low quality recording on a linear track on the edge of the tape; this is usually mono but some machines provide stereo, sometimes with Dolby noise reduction. Additionally many machines record in stereo, using rotating heads within the video signal. The actual sound uses an FM carrier with a noise reduction, using high frequency pre-emphasis and 3 : 1 compression.

WAV   The .WAV format originated from Microsoft as a simple format for storing audio for games, etc. The format has been expanded for professional use adding text fields, etc. If an exported .WAV file begins with a click on another piece of software, then it is possible that this software is not reading the file correctly. There is usually an option to export the files without the text information, which can solve the problem. There is also an ADPCM version of this format giving 4 : 1 lossy compression. This is best regarded as an end-user format.

WMA (Windows Media Audio)   A Microsoft rival to MP3 files for streaming over the Internet.

Wow   Slow variation of pitch, traditionally caused by an off-centre gramophone record or a slipping capstan pulley on a tape machine. Digital audio systems are entirely free of this, unless it is deliberately introduced as a special effect.

XLR   Connector widely used in the professional audio field.

XY stereo   In the context of microphones, this is sometimes used to indicate the use of a coincident pair of microphones as opposed to spaced (AB).

Zip drive   Proprietary form of removable hard disk cartridge.

Zip format   Lossless data compression format much used to reduce the size of program and data files on disk and sent via modems. Unfortunately the assumptions that it makes do not apply to audio, and files can end up larger after Zip compression than they were before.

Index

Focal Press

www.focalpress.com
Join Focal Press on-line
As a member you will enjoy the following benefits:

- an email bulletin with **information on new books**

- a regular **Focal Press Newsletter**:
    - featuring a selection of new titles
    - keeps you informed of **special offers, discounts and freebies**
    - alerts you to **Focal Press news and events** such as author signings and seminars

- complete access to **free content** and reference material on the focalpress site, such as the focalXtra articles and commentary from our authors

- a **Sneak Preview** of selected titles (sample chapters) *before* they publish

- a chance to have your say on our **discussion boards** and **review books** for other Focal readers

Focal Club Members are invited to give us feedback on our products and services.
Email: worldmarketing@focalpress.com – we want to hear your views!

Membership is **FREE**. To join, visit our website and register. If you require any further information regarding the on-line club please contact:

> Lucy Lomas-Walker
> Email: l.lomas@elsevier.com
> Tel: +44 (0) 1865 314438
> Fax: +44 (0)1865 314572
> Address: Focal Press, Linacre House,
> Jordan Hill, Oxford, UK, OX2 8DP

Catalogue

For information on all Focal Press titles, our full catalogue is available online at www.focalpress.com and all titles can be purchased here via secure online ordering, or contact us for a free printed version:

USA
Email: christine.degon@bhusa.com
Tel: +1 781 904 2607 T

Europe and rest of world
Email: j.blackford@elsevier.com
el: +44 (0)1865 314220

Potential authors

If you have an idea for a book, please get in touch:

USA
editors@focalpress.com

Europe and rest of world
focal.press@repp.co.uk

CHESTER COLLEGE LIBRARY